Jane's

Space
Recognition Guide

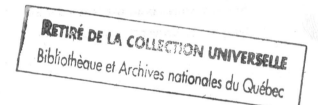

Jane's Recognition Guides

Aircraft
Airline
Guns
Special Forces
Tank
Train
US Military Aircraft
Warship

Space
Recognition Guide

Peter Bond

Collins

First published in 2008 by **Collins**

HarperCollinsPublishers
77-85 Fulham Palace Road
Hammersmith
London W6 8JB
UK
www.collins.co.uk

HarperCollins Publishers Inc
10 East 53rd Street
New York
NY 10022
USA
www.harpercollins.com

www.janes.com

ISBN 13 978-0-00-723296-3

Layout: Susie Bell
Editorial: Louise Stanley, Jodi Simpson

Printed and bound by
Printing Express, Hong Kong

Contents

List of Entries

Major Space Agencies

Selected Historic Missions

Historic Launchers

Current/Future Launchers

Launch Sites

Civil Communications & Applications Satellites

Military Satellites

Scientific Satellites: Astronomy

Earth Remote Sensing Satellites

Human Spaceflight

Futures

Glossary

Fifty Years of Spaceflight

Throughout its existence, the human race has been driven by a desire to explore. Some 70,000 years ago, the first representatives of *Homo sapiens* began to migrate from their homeland in Africa. Our ancestors overcame advancing ice sheets, competition for resources, and intimidating physical barriers before eventually exploring and settling in every continent. Today, there are few corners of the globe where humanity has not left its mark.

There are many motivations for exploration. Often, population pressure necessitated a search for virgin land and untapped natural resources. However, always in the background was a curiosity to find out what lay in the next valley or on the other side of the ocean, and a search for a deeper understanding of our surroundings.

Today, with the major exception of the deep ocean floors, our planet has largely been explored and surveyed. However, the spirit of past explorers lives on in the desire to learn more about humanity's place in the universe, something that can only be achieved by venturing forth to visit and study new worlds.

For the first time in history, technology has reached a state of sophistication that enables us to compare the evolution of our planet with its neighboring worlds, to catalogue the resources of these worlds, and to search for the answers to such fundamental questions as "How did life evolve?", "Why is the Earth so different from its neighbors?", and "Are we alone?"

The Birth of the Space Age

Inevitably, concerns over personal and national security have always played a part in humanity's inexorable drive to settle and colonize new territories. Our first, faltering footsteps towards the Final Frontier were no exception. The early years of the Space Age resonated with the rhetoric of the cold war between the United States and the Soviet Union.

Fifty years ago, the launch of the first artificial satellite marked the beginning of a superpower race, in which national prestige and technological supremacy were the major motivations. As the beach-ball-sized Sputnik 1 circled the globe, transmitting a monotonous beeping signal, the citizens of Western nations looked in panic towards the heavens and implored their leaders to protect them from the apparent Communist threat.

This East–West rivalry was the most significant motivation behind President John F. Kennedy's 1961 rallying call to land a man on the Moon before the decade was out and return him safely to the Earth. It is a sobering thought that the $25 billion Apollo program, which inspired a whole generation of scientists and engineers and—at least at the beginning—enthralled millions of people around the world, would not have taken place without the overwhelming political desire to demonstrate American prowess and beat the Soviets.

The early years of the Space Age, starting with the launch of Sputnik in October 1957, were marked by many setbacks and failures. Hidden behind the Iron Curtain, the Soviets were able to hide their disasters, whereas the American "flopniks" took place in the full glare of the world's media. Soviet cosmonauts and spacecraft designers were airbrushed out of photographs and their anonymity was guaranteed. Even the fact that Yuri Gagarin ejected from his Vostok capsule and parachuted separately to Earth was deliberately withheld. Only with the gradual opening up of the Soviet program many years later were the mysterious chief designers, the unknown cosmonauts, and the numerous launch failures exposed for all to see.

Although the first launches and missions involved a great deal of trial and error, the pace of progress was rapid. Once the ability of animals to cope with the stresses and hazards of spaceflight was confirmed, astronauts and cosmonauts learned not only how to survive in this unearthly environment, but also how to utilize its unique properties for scientific research.

In April 1961, Gagarin became the first human to observe the blue Earth shining against the black velvet of space. In 1965, Alexei Leonov became the first person to walk in the vacuum of space. The following year, Neil Armstrong and David Scott survived a near-disaster to achieve the first space docking. Their courageous exploits helped to prepare the way for Armstrong and Buzz Aldrin to set foot on the Mare Tranquillitatis in July 1969—thus picking up the gauntlet thrown down by Kennedy only eight years earlier. Eventually, nine manned missions flew to the Moon, and twelve intrepid explorers left their footprints on the dust-covered Moon.

After the dramatic successes of the Apollo program, humanity retreated to near-Earth space, where we have remained for 35 years. The launches of the first orbital space stations and the advent of the reusable Space Shuttle led to an increasing emphasis on microgravity research and the commercialization of low Earth orbit. At the same time, increasingly advanced, automated spacecraft brought about a revolution in such fields as telecommunications, weather forecasting, Earth remote sensing, and navigation.

Staggering Progress

Since Sputnik shook the world, more than 6500 satellites have been launched. Although most of these have been inserted into various orbits around the Earth, several hundred robotic ambassadors have been despatched throughout the Solar System. The first, fleeting surveys undertaken by flyby spacecraft were soon followed by long-lived orbiters that were able to map a planet's entire surface.

In recent years, these surveyor missions have been increasingly supplemented by autonomous rovers and suicidal atmospheric probes. Every planet has been visited at least once, with Comet Churyumov-Gerasimenko and the dwarf planets Ceres and Pluto soon to be added to the list. Even more exciting is the potential to bring back samples of alien material, as originally demonstrated by the Apollo astronauts and the Soviet automated Luna landers. Samples of solar wind and comet dust have already been collected by spacecraft, and Japan's Hayabusa is currently struggling to return home with a possible payload of a few grains of asteroid soil.

Just as staggering is the progress that has been made in space-based astronomy. Refurbished by visiting astronauts on four occasions since its launch in 1990, the Hubble Space Telescope has continued to amaze with its spectacular and colorful images. Every wavelength in the electromagnetic spectrum is now utilized by state-of-the-art observatories, with the result that a menagerie of exotic objects, from extrasolar planets to supernovae and black holes, is now open to scrutiny. Even more ambitious future missions are being designed to try to prove the existence of invisible dark matter, dark energy, and gravity waves.

The peaceful uses of space are many and varied, but, like a large iceberg in the ocean, the global space program has a less familiar, hidden side. Although the fears of the Sputnik generation that space would become a battleground have not been fulfilled, the Earth is girdled by dozens of military satellites designed to link armed forces around the globe, warn of ballistic missile launches, and spy on rival nations. Although they rarely grab the headlines, huge sums are spent on these secret programs. Occasionally, as is the case with the US military Global Positioning System, a program may eventually benefit society as a whole.

A Multinational Endeavor

Since the Space Age began, the United States has been the pacesetter for space expenditure, and this trend is likely to continue for the foreseeable future. Even so, the $16.3 billion annual budget of NASA is now quite modest compared with its heyday in the 1960s, when Apollo accounted for 4% of the federal budget. At the same time, the US spends about $20 billion per year on military space activities.

Second in the space expenditure stakes, but a long way behind the US military and NASA, is the 17-nation European Space Agency (ESA). Europe has been involved in human space exploration since the early 1970s, when NASA invited participation in its crewed shuttle program. The European nations opted to develop a modular research facility, named Spacelab, that would operate in the orbiter's payload bay.

In 1983, during the maiden flight of Spacelab, German Ulf Merbold made history by becoming the first ESA astronaut to fly in space and the first non-American "guest" to be launched onboard a US space vehicle. Since 1978, more than 30 astronauts from ESA member states have flown on 48 missions. Of these, 27 were cooperative programs with NASA, while 21 involved collaboration with the former Soviet Union.

Such multinational cooperation has become a theme of recent years as many of the cold war rivalries have dissipated. Most significantly, 16 countries have come together to work as partners on the International Space Station (ISS), the largest, most complex, and most expensive spacecraft ever to orbit the Earth.

European contributions to the ISS program include the Columbus laboratory, at least half a dozen Automated Transfer Vehicles, the European Robotic Arm, three Multipurpose Logistics Modules, two ISS Nodes, the Data Management System for the Russian segment, and a European-built observation module called a cupola. In addition to these elements, Europe provides specialist scientific facilities, including a Microgravity Glovebox and various refrigerators and freezers. Another major contributor to the ISS is Russia, which has provided several modules (one funded by the US) and a reliable, low-cost space transportation system involving its Soyuz and Progress craft. Like Europe, Japan relies on the US and Russia to deliver its astronauts to orbit, but its scientists are looking forward to conducting some groundbreaking microgravity research in the new Kibo laboratory. A Japanese cargo transfer vehicle, the HTV, is also under development. Meanwhile, Canada has created its own specialized niche by developing a series of advanced robotic arms and manipulators.

At the same time, the US, Russia, Europe, Japan, China, and India have developed their own autonomous access to space, while Brazil has ambitions to do the same. Since the first Ariane launch in December 1979, the family of European launch vehicles has been the strategic key to the development of all European space applications, dominating the world's commercial launch market for many years.

Today, the heavy-lift Ariane 5 can carry a payload of two large commercial satellites to geostationary transfer orbit, and a modified version will soon launch the first ATV on a rendezvous-and-docking mission to the ISS. European launch capabilities will also be augmented with the introduction of the small Vega vehicle and a launch pad for the Russian Soyuz rocket at Europe's spaceport in Kourou, French Guiana.

In the current modest launch market, competition for launch contracts is high, even though China's highly reliable Long March vehicles and India's PSLV and GSLV launchers are largely excluded from the commercial market by restrictive US technology-transfer regulations. The oversupply of launchers is not helped by the availability of decommissioned missiles that can deliver small payloads to low Earth orbit. Russia, in particular, retains a sizeable legacy of launchers and modified missiles left over from the cold war era.

However, it is a modern communist power that has taken over the secretive role of the former Soviet Union. Although it is no longer possible to hide a launch disaster from the watching world media, the Chinese space program, largely overseen by the People's Liberation Army, remains reluctant to open up its secrets to the outside world.

What is clear, however, is that society has been irrevocably changed by the Space Age. The lives of just about everyone in the developed world are touched in some way by space technology and innovation—whether it be weather satellites, Earth-imaging spacecraft, hand-held navigation systems and mobile phones, satellite TV broadcasting, or broadband Internet connections. Even in the less developed countries, space is seen as a driver for high-tech industries and as an inspiration for young scientists and engineers. There seems little doubt that the advances of the last 50 years will be more than matched by the innovations of the next five decades.

Given the many varied and different missions and types of technology associated with space over the past fifty years we have been unable to include every single rocket and spacecraft—however the author and editorial team have carefully selected the most significant or recent missions/pieces of hardware, and tried to give as representative a spread of the subject as possible. In addition, we hope that the photographs are sufficiently representative of the subjects covered here; the reader should bear in mind that in some cases photos are very rare or have restricted access – so some compromises have necessarily been made.

Peter Bond

Picture Credits

AEB 130, 153; AIP 138; Arianespace 177; Alcatel 169; Alcatel Lucent 46; ASI 246; Astra 173; BAE 73; Ball Aerospace 231, 270; Bigelow Aerospace 151, 185; Birmingham University 328; Boeing (Hughes Space Communication) 211; Boeing/Thomas Baur 105; Boeing 48, 55, 172, 174, 192, 206, 207, 208, 209, 210, 212, 213, 230, 240; Brian Harvey 372; CfA Harvard 267; CGWC 111, 162; CGWIC 152; CIA 43; CNES 83, 239, 249, 345; CNSA 24; CSA 23, 262, 356; Digital Globe 242; DLR 360; EADS-Astrium 82, 168, 170, 180, 191, 193, 195, 203, 204, 224, 236, 320, 336, 339, 358; Energia 40, 41, 51, 72, 75, 84, 89; 94; ESA 25, 26, 27, 71, 76, 80, 85, 114, 123, 171, 182, 186, 254, 256, 257, 258, 261, 263, 271, 275, 278, 279, 282, 289, 293, 304, 306, 313, 329, 331, 333, 351, 357, 380; ESA/S. Corvaja 118; ESA-CNES-Arianespace 102, 140; ESA-D Ducros 314, 368; ESA-NASA 307; Eurockot 116, 146; GeoEye 136; Hampton Univ 318, 324; Hispasat 190; IAI 117, 228, 332; INPE 326; Inter-kosmos-CNES 64, 227; Iridium 196; ISRO 28, 29, 108, 115, 150, 194, 277, 325, 344, 346; Izmiran 347; Jane's 143, 335; Japanese Ministry of Land Infrastructure and Transport 200; JAXA 30, 31, 86, 87, 88, 109, 154, 156, 178, 247, 266, 288, 319, 340, 369; JAXA/Akihiro Ikeshita 287, 305; JHU 361; JHU-APL 225, 300; KARI 348; Khrunichev 100; Kosmotras 106; LASP 341; Lavochkin 57, 58; Lockheed Martin 166, 184, 197, 208, 216, 219, 220, 222, 226, 234, 253, 312, 342; Makeyev-Plan Soc 124; Mark Wade, Encyclopedia Astronautica 217, 232; Mid-Atlantic Spaceport 160, 161; MIT 255; Nahuelsat 201; NASA 34, 42, 45, 49, 50, 52, 54, 56, 61, 63, 65, 66, 69, 74, 77, 91, 92, 96, 104, 248, 251, 252, 259, 260, 265, 268, 269, 274, 280, 281, 283, 285, 290, 291, 292, 294, 295, 296, 297, 298, 299, 301, 302, 303, 310, 311, 315, 321, 322, 323, 327, 338, 343, 349, 352, 359, 362, 364, 365, 371, 373, 379; NASA/Bill Ingalls 22; NASA-Ames 67, 68, 284; NASA-ESA 276; NASA-GSFC 264, 334, 353; NASA-HQ-GRIN 32; NASA-JPL 47; NASA-KSC 103, 128, 250, 309, 375; NASA-LaRC 93; NASA-MSFC 33, 38, 39, 62, 90, 95, 97, 101, 286, 308, 370; NASA-NSSDC 60; Netherlands Space Research Institute 70; New Mexico State Univ 148; Northrup Grumman 237; Novosti 59; NPO Mashinostroveniva 120; NPO PM 181, 238; OHB 110, 233; OSC (Orbital Sciences) 112, 113, 121, 175, 183, 199, 202, 205, 337, 354; Peter Bond 53, 188; Qinetiq 363; Roscosmos 122; RSF 223; SAFT 350; Scaled Composites 374; Sea Launch 125, 144; Selenia Spazio 235; SpaceX 107, 142; SS/Loral 179, 243; SSTL 330; Starsem 132, 187; Swedish Space Corp. 119; Teledyne Brown Engineering 189, 229; Thales Alenia 176, 198, 218; USAF 81, 134, 158, 221, 241; Verhaert/ESA 355; Wikipedia 44

Major
Space
Agencies

Brazilian Space Agency (AGÊNCIA ESPACIAL BRASILEIRA/AEB)

Marcos Pontes pictured

AEB was created on February 10, 1994, as an autonomous agency under the Ministry of Science and Technology. It replaced the Brazilian Commission for Space Activities (COBAE), which led Brazil's space activities from the 1970s. AEB's headquarters are in Brasília. The agency comprises four directorates: Space Policy and Strategic Investment; Satellite Applications and Development; Space Transportation and Licensing; and Planning, Budget, and Administration. Brazil's Ministry of Defense operates the Alcântara Launch Center, the site from which AEB tests its small satellite orbital launch vehicle, the Veículo Lançador de Satélites (VLS).

The national space budget in 2003 was $56 million. Traditionally, the country's satellite activity has concentrated upon Earth observation applications, including remote sensing, meteorology, and oceanography. Other priorities are telecommunications science and the development of an indigenous launch capability.

Brazil favors a policy of joint technological development with other, more advanced nations. Initially relying heavily on the United States, the agency ran into difficulties over technology transfer restrictions. Recently, Brazil has extended its cooperation to other countries. In 2003, Brazil agreed to upgrade Alcântara for international launches of Ukraine's Tsyklon-4 rocket. The Russian Federal Space Agency has agreed to assist in the development of a new launch vehicle family. A number of satellites have been developed under the China-Brazil Earth Resources Satellite (CBERS) program. Involvement in the International Space Station (ISS) has suffered from financial cutbacks, but a Brazilian astronaut flew to the ISS in a Soyuz spacecraft in March 2006.

Canadian Space Agency (L'AGENCE SPATIALE CANADIENNE/CSA)

John H. Chapman Space Center, Canada pictured

The CSA was established by the Canadian Space Agency Act of March 1989, which came into force on December 14, 1990. The legislated mandate of the CSA is: "To promote the peaceful use and development of space, to advance the knowledge of space through science and to ensure that space science and technology provide social and economic benefits for Canadians."

The CEO of the agency is the president, whose rank is equivalent to that of a deputy minister and who reports to the Minister of Industry. The Agency has five core areas: Space Programs, Space Technologies, Space Science, the Canadian Astronaut Office, and Space Operations.

The CSA has about 635 employees and some 170 service contractors, 90% of whom work at the John H. Chapman Space Center, the agency headquarters in Longueuil, Quebec. The Agency's annual budget is around US $240 million.

The CSA has several partnerships and collaborative programs with space agencies in other countries, such as NASA, ESA, and JAXA. Since January 1, 1979, Canada has had the special status of a cooperating state with ESA, enabling it to participate in ESA discussions, programs, and activities. Most Canadian space activities are conducted in collaboration with international partners. Canada's major technological contributions include the Canadarm mechanical arm on the Space Shuttle, together with the Canadarm2 and the Mobile Servicing System on the International Space Station. Eight Canadian astronauts have flown into orbit on the Space Shuttle.

China National Space Administration (CNSA)

CNSA launch (below); CNSA–ESA cooperation (opposite)

Until 1998, China's space program was controlled by the People's Liberation Army (PLA). In that year, the civil and military aspects of the program were separated with the establishment of the China National Space Administration.

The CNSA reports to the Commission of Science, Technology and Industry for National Defense. It is responsible for signing governmental agreements in the space area, including intergovernmental scientific and technical exchanges; for enforcing national space policies, as laid down in each five-year plan; and for managing space science, technology, and industry. This includes responsibility for the planning and promotion of civilian space activities, such as the manufacture of launchers and civilian satellites. There are four departments under the CNSA: General Planning; System Engineering; Science; Technology, and Quality Control; and Foreign Affairs.

In recent years, China has shown an increasing interest in collaborating with international partners. Among those countries to have signed space cooperation agreements with them are Brazil, Chile, France, Germany, India, Italy, Pakistan, Russia, Ukraine, the United Kingdom, and the United States. However, commercial launch activities have been severely curtailed by US trade restrictions intended to prevent arms proliferation and transfer of sensitive technologies.

China has developed a variety of telecommunications, Earth resources, and meteorological satellites, although many have been dual-use in nature, supporting both civil and military efforts. National space spending is estimated at around $3 billion per year. In the past decade, China's investment in space science, including infrastructure and programs, exceeded $112.5 million. Key research areas will also include astronomy and solar physics, space physics and solar system exploration, microgravity sciences, and space life science. Following on from the Double Star collaboration with ESA, CNSA plans to launch three satellites to conduct exploration of the Sun and solar wind. Another priority is deep space exploration, focusing on lunar and Mars exploration. The lunar program will have three stages: the Chang'e-1 lunar orbiter, a surface rover, and a sample return mission.

The PLA retains control of the space infrastructure, such as the launch centers and mission control centers.

European Space Agency (ESA)

ESA headquarters, Paris (below); ESA headquarters interior (right)

The European Space Agency was established on May 31, 1975, by merging the activities of the European Launcher Development Organization (ELDO) and the European Space Research Organization (ESRO).

Dedicated to the peaceful exploration of space, the original agency had 10 founding members: Belgium, West Germany, Denmark, France, the United Kingdom, Italy, the Netherlands, Sweden, Switzerland and Spain. More recent additions to the membership are Austria, Ireland, Portugal, Greece, Norway, Finland, and Luxembourg, bringing the number of member states to 17. Canada, Hungary, Poland, and the Czech Republic also participate in some projects under cooperation agreements.

ESA has no formal link to the European Community (EC). However, in recent years the ties between them have been reinforced by the increasing role that space plays in strengthening Europe's global political and economic role, and in supporting European policies through space exploration and utilization. Recent joint initiatives include the Galileo global navigation satellite system and Global Monitoring for Environment and Security program, which was endorsed by ESA's Ministerial Council in December 2005.

ESA has a staff (excluding subcontractors and national space agencies) of about 1900, with an annual budget of about $4 billion in 2006. Most of the staff are based at the European Space Research and Technology Center (ESTEC) in the Netherlands. The agency operates on the basis of geographical return: that is, it invests in each member state, through industrial contracts for space programs, an amount more or less equivalent to each country's financial contribution.

ESA's headquarters are in Paris, France, where policies and programs are decided upon. ESA also has centers in various European countries: European astronauts are trained for future missions at the European Astronauts Center (EAC) in Cologne, Germany; the European Space Astronomy Center (ESAC) in Villafranca del Castillo, Spain, hosts the science operation centers for all ESA astronomy and planetary missions together with their science archives; the European Space Operations Center (ESOC) in Darmstadt, Germany, is responsible for controlling ESA satellites in orbit; the ESA Center for Earth Observation (ESRIN) in Frascati, Italy, is responsible for collecting, storing, and distributing Earth observation satellite data and is also the agency's information technology center; the ESTEC in Noordwijk, the Netherlands, is the main center for ESA spacecraft design, technology development, and testing. ESA has liaison offices in Belgium, the United States, and Russia.

ESA's spaceport in Kourou, French Guiana, is owned by the French national space agency, Centre National d'Études Spatiales (CNES). It is used by ESA and the commercial space launch company Arianespace to deliver European payloads to space.

ESA has developed the Columbus laboratory and other hardware for the International Space Station (ISS), and its Ariane rockets have become world leaders in the commercial launch market. ESA astronauts have flown on the Space Shuttle and Soyuz to Mir and the ISS. The agency also has highly successful space science and Earth observation programs. ESA collaborates with all other space powers, most recently with China on the Double Star program.

Indian Space Research Organization (ISRO)

Satish Dhawan Space Center (below); INSAT-4c satellite (opposite)

ISRO was established in 1969 as India's primary space R&D organization, responsible for developing launcher and propulsion systems, launch sites, and satellites and their tracking networks. Personnel totals rose from 13,488 in 1986 to 16,800 in 1996 and stood at 16,400 in 2002. There are numerous centers and units. The Vikram Sarabhai Space Center near Trivandrum, the base for launcher and propulsion development, is ISRO's largest facility with 4700 personnel. Another major center is the Liquid Propulsion Systems

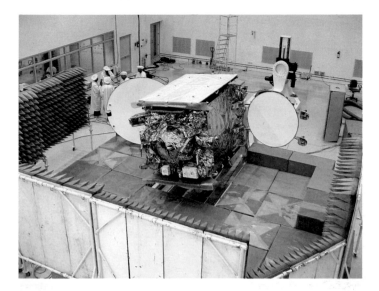

Center which has development wings in Bangalore and Trivandrum, supported by major test facilities at Mahendragiri for a wide spectrum of liquid motors, from reaction control system thrusters to the Vikas and cryogenic engines.

ISRO's annual budget in 2005–6 was about $550 million, with large increases expected in future years. Its prime objective is to develop space technologies and their application to various national tasks. It has created two major space systems: INSAT satellites for communication, television broadcasting ,and meteorological services; and the Indian Remote Sensing (IRS) satellites for monitoring and management of Earth resources. It has also developed the PSLV and GSLV rockets to launch national and international payloads. It is now beginning to consider deep space missions and a human spaceflight program.

ISRO works with the Antrix Corporation, the commercial arm of the Department of Space, to promote commercialization of products and services from the Indian space program.

Japan Aerospace Exploration Agency (JAXA)

Sagamihara Center (below); Tsukuba Center (right)

JAXA was formed on October 1, 2003, through the integration of three space-related organizations: the Institute of Space and Astronautical Science (ISAS), the National Aerospace Laboratory (NAL), and the National Space Development Agency (NASDA). JAXA's headquarters are in Tokyo. The agency also maintains over 20 field centers and tracking station sites in Japan, as well as offices in Washington DC; Houston, Texas; Paris, France; Bangkok, Thailand; and a liaison office at NASA's Kennedy Space Center, Florida. JAXA has a regular staff of about 1700 plus graduate students and foreign researchers.

JAXA is organized into six key divisions and institutes: the Office of Spaceflight and Operations; the Office of Space Applications; the Institute of Aerospace Technology (IAT); the Institute of Space and Astronautical Science (ISAS); the Human Space Systems and Utilization Program Group; and the Aviation Program Group (APG).

ISAS retained its name after JAXA was formed and continues to focus upon space science, observation, and planetary research. It was responsible for launching Ohsumi, Japan's first satellite, in 1970. It is currently the lead organization for the Hayabusa asteroid sample return mission and has plans to send spacecraft to Venus and Mercury in the coming years. ISAS works in cooperation with Japanese universities wishing to conduct space science research.

The Human Space Systems and Utilization Program Group deals with all human spaceflight activities, including astronaut training and International Space Station (ISS) activities. Japan is providing the Kibo laboratory for the

ISS and several Japanese astronauts have flown to the station on the Space Shuttle.

APG leads aviation research and continues to build on NAL's past achievements. Projects include the Hypersonic Experimental Vehicle, a passenger plane similar to the Concorde, and Unmanned Aerial Vehicle development.

JAXA's budget declined from a high of ¥226 billion ($1.9 billion) in 1999 to ¥176 billion ($1.5 billion) in 2005. This decline caused delays in a number of rocket and satellite programs. However, the 2006 budget saw a modest increase to ¥180.1 billion ($1.53 billion).

In April 2005, JAXA released its 20-year vision statement. This included the development of "launch vehicles and satellites with the highest reliability and world-class capability, with a view to building a secure and prosperous society. JAXA shall also work toward taking a leading position in the world in space science and begin preparations for Japan's own human space activities and for the utilization of the Moon. JAXA shall further conduct a flight demonstration of a prototype hypersonic aircraft that can fly at the speed of Mach 5."

One key initiative is Sentinel Asia, a rapid-response disaster management system, based on satellite data and images, that involves 19 nations. In addition to its leading role in space astronomy, the agency plans to send probes to Mercury, Venus, and Jupiter and to conduct a program of robotic lunar exploration.

National Aeronautics and Space Administration (NASA)

The Jet Propulsion Laboratory, California (below); ISS Payload Operations Center, MSFC (right)

NASA was formally established on October 1, 1958, replacing the National Advisory Committee for Aeronautics. Its remit was to plan and execute the US civil space program, which was by then in competition with the Soviet Union, following the launches of the first Sputniks.

Today, NASA comprises four Mission Directorates: Exploration Systems, whose main objective is to support the Vision for Space Exploration, involving sending humans to the Moon, Mars, and beyond; Space Operations, charged with the management of the International Space Station and the Space Shuttle programs, as well as launch services and space communications; Science, responsible for space science research including solar system exploration and Earth observation; and, last, Aeronautics Research.

NASA employs about 18,000 civil servants, over half of whom are scientists and engineers. Roughly 1100 personnel are located at NASA Headquarters in Washington, DC. The Agency also operates ten major research centers and facilities. These include the John H. Glenn Research Center at Lewis Field, Cleveland, Ohio, which is NASA's lead organization for aeronautics and space propulsion, and microgravity science research; the Johnson Space Center in Houston, Texas, NASA's lead center for human space flight and home to its astronaut corps; and the Jet Propulsion Laboratory (JPL) in Pasadena, California, a federally funded facility operated by the California Institute of Technology. JPL has been under contract to NASA since 1958. Its main areas of expertise are planetary exploration and

Earth science research. Although NASA is its main client, JPL also performs work for other US government agencies, such as DOD.

The Kennedy Space Center in Florida was the launch site for the Apollo and Skylab missions. It is currently responsible for processing, launching, and landing the Space Shuttle and its payloads, including International Space Station (ISS) components.

NASA's budget peaked in 1966, prior to the Apollo Moon landings, when it accounted for 4% of the total US federal budget. In 2006 and 2007, its overall budget was about $16.2 billion (about 0.6% of the federal budget). More than $3 billion is devoted to Exploration Systems, which plans to develop a Crew Exploration Vehicle (CEV) and Crew Launch Vehicle in order to complete a human lunar surface expedition no later than 2020. Human spaceflight, including flying the Space Shuttle and completion of the ISS by 2010, is allocated $7 billion. After retirement of the shuttle in 2010, crews will be carried to the ISS by Soyuz and the new CEV. The budget also provides $5.2 billion in funding for NASA's science missions and $900 million for aeronautics research.

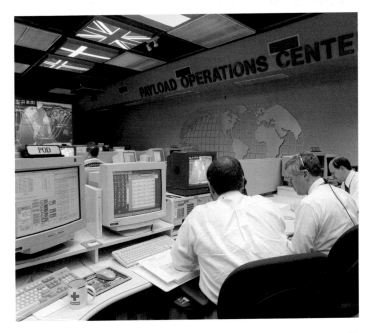

Russian Federal Space Agency (RKA/ROSKOSMOS)

Mission control centre pictured

The Russian Federal Space Agency, formerly the Russian Aviation and Space Agency (Rosaviakosmos), is the government agency responsible for Russia's civil space program. It was created by Presidential decree in February 1992, after the breakup of the former Soviet Union and the dissolution of the Soviet space program. RKA headquarters are located in Moscow, where it has about 300 employees, with another 600 offsite.

RKA is responsible for all manned and unmanned nonmilitary space flights and uses the Russian technology companies, infrastructure, and launch sites that belonged to the former Soviet space program. Its principal areas of responsibility include state programs, human spaceflight, science and commercial satellites, international affairs, and ground infrastructure.

The major launch center at Baikonur, formerly part of Soviet territory, is now leased from the government of Kazakhstan for $115 million per year. In June 2005, the lease was extended until 2050. The former military base is scheduled to be handed over to RKA control in 2007. RKA and the military

also share control of the Yuri Gagarin Cosmonaut Training Center at Zvezdny Gorodok (Star City), near Moscow.

RKA manages more commercial launches per year than any other agency or country. It has a 25% stake in the Starsem launch company and recently signed an agreement with ESA to launch Soyuz vehicles from Kourou, French Guiana.

Since its creation, the agency has experienced a constant lack of funding, which has curtailed development of new spacecraft and launch vehicles. Science funding, in particular, has suffered, with very few new missions launched. This has resulted in increased collaboration with international partners; efforts to encourage launches of commercial satellites, including the marketing of various missile systems as commercial launchers; and space tourism.

As a major partner in the International Space Station (ISS), Russia has provided three modules, one under a commercial arrangement with NASA, as well as regular Soyuz and Progress missions to the station. After the initial ISS contract with NASA expired, RKA agreed to sell seats on Soyuz spacecraft for approximately $21 million per person each way and to provide Progress transport flights ($50 million each) until 2011.

The recent boom in oil and gas prices has led to promises of increased funds, with a 2006 budget of about $900 million, a 33% increase over the 2005 budget. Under the current 10-year plan, the RKA budget will increase 5–10% per year. This may allow further expansion of the Russian section of the ISS, the development of the Kliper reusable spacecraft, and the first launches of spacecraft to Mars and the Moon in many years.

Selected Historic Missions

Sputnik 1 (PS-1) USSR

First orbital artificial satellite

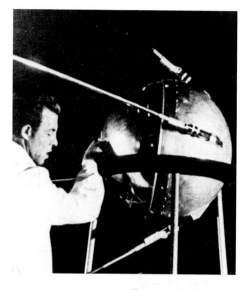

SPECIFICATION:

MANUFACTURER: OKB-1 (Korolev)
LAUNCH DATE: October 4, 1957
ORBIT: 947 x 228 km (588 x
142 mi), 65° inclination
LAUNCH SITE: Baikonur,
Kazakhstan
LAUNCHER: Sputnik (R-7)
LAUNCH MASS: 83.6 kg (184 lb)
BODY DIAMETER: 0.6 m (1.9 ft)
PAYLOAD: 2 radio transmitters

Sputnik 1 enabled the USSR to launch the world's first artificial satellite. The satellite was a smooth aluminum sphere about the size of a basketball that contained a radio transmitter and batteries. Trailing behind it were four whip aerials of between 1.5 and 2.9 m (4.9 and 9.5 ft) in length.

Simple Satellite 1 (PS-1) was launched on a slightly modified R-7 ICBM. The changes included removing the radio system from the core booster and modifying the firing regime of both the core and strap-on engines.

After launch, it was named "Sputnik," meaning "Companion" or "Satellite." The sphere was filled with nitrogen under pressure. As it circled the Earth every 96 min, its "beeping" signal delivered a message of apparent Soviet military superiority in space that caused major concern among Western nations. Spacecraft tracking provided data about the density of the upper atmosphere. Sputnik transmitted for 21 days but survived in orbit for 92 days before burning up during reentry on January 4, 1958.

Explorer 1 USA

First US satellite to orbit earth

SPECIFICATION:

MANUFACTURER: Jet Propulsion Laboratory
LAUNCH DATE: January 31, 1958
ORBIT: 360 x 2532 km (223 x 1572 mi), 33.3° inclination
LAUNCH SITE: Cape Canaveral, Florida
LAUNCHER: Jupiter C / Juno 1
LAUNCH MASS: 14 kg (30.8 lb) including integral upper stage motor
DIMENSIONS: 2 x 0.2 m (6.6 x 0.5 ft) including attached rocket stage
PAYLOAD: Geiger counter; micro-meteorite detectors; internal and external temperature sensors

After the launch of Sputnik 1, the US Army Ballistic Missile Agency was directed to launch a satellite using the Jupiter C rocket developed by Wernher von Braun's team. The Jet Propulsion Laboratory designed and built the satellite payload in less than three months.

Explorer 1 carried 8 kg (17.6 lb) of instruments designed to collect data on cosmic rays, meteorites, and temperature. The cosmic ray detector was designed to measure the Earth's radiation environmen and revealed a much lower cosmic ray count than expected. Dr. James Van Allen theorized that the instrument may have been saturated with strong radiation from a belt of charged particles trapped in space by Earth's magnetic field. The existence of these radiation belts was later confirmed, and they became known as the Van Allen belts.

Explorer 1 followed an elliptical orbit with a period of 114.8 minutes. The satellite itself was 2 m (80 in) long and 15.9 cm (6.3 in) in diameter. Its final transmission came on May 23, 1958. It burned up in Earth's atmosphere on March 31, 1970, after more than 58,000 orbits.

Sputnik 2 (PS-2) USSR

First satellite to carry a live animal

SPECIFICATION:

MANUFACTURER: OKB-1 (Korolev)
LAUNCH DATE: November 3, 1957
ORBIT: 225 x 1671 km (140 x
1038 mi), 65.3° inclination
LAUNCH SITE: Baikonur,
Kazakhstan
LAUNCHER: Sputnik (R-7)
LAUNCH MASS: 508 kg (1118 lb)
BODY DIMENSIONS: 1.2 m (3.9 ft)
long
PAYLOAD: Pressurized container
with dog Laika; Geiger counters;
2 ultraviolet and x-ray spectro-
photometers

The first satellite to carry a live animal was the result of Premier Khrushchev's "request" to launch a satellite in time for the 40th anniversary of the Great October Revolution in November 1957. Sputnik 3 was not ready, so, with less than four weeks to prepare, Sergei Korolev's OKB-1 design team decided to modify the already-built PS-2 "Simple Satellite" by adding a container derived from the payload of the R-2A suborbital scientific rocket.

The spherical, pressurized container carried the dog Laika, the first living creature to orbit the Earth. The padded cabin provided enough room to lie down or stand. Laika was fitted with electrodes to monitor life signs.

The scientific purpose of the mission was to gather data on the effects of weightlessness on an animal, in preparation for a human flight. The satellite was not designed to return to Earth with its occupant. Laika survived in orbit for perhaps 5–7 h (figures vary) but died when the cabin overheated. Sputnik 2 reentered the atmosphere on April 14, 1958, after 2570 orbits.

Luna 3 (Automatic Interplanetary Station/Lunik 3) USSR

First images of the Moon's far side

SPECIFICATION:

MANUFACTURER: OKB-1 (Korolev)
LAUNCH DATE: October 4, 1959
ORBIT: 40,300 x 476,500 km
(25,041 x 296,091 mi), 73.8°
inclination
LAUNCH SITE: Baikonur,
Kazakhstan
LAUNCHER: Luna 8K72 (R-7
Sputnik with Block E escape
stage)
LAUNCH MASS: 278.5 kg (615 lb)
DIMENSIONS: 1.3 x 1.2 m (4.3 x
3.9 ft)
PAYLOAD: Yenisey-2 TV imaging
system; micrometeoroid
detector; cosmic ray detector

In January 1959, Luna 1 became the first spacecraft to fly past the Moon. In September, the USSR successfully crashed the Luna 2 spacecraft on the Moon after five failed attempts. It was soon followed by Luna 3, the first probe to send back images of the Moon's far side. Luna 3 was a roughly cylindrical spacecraft powered by solar cells, with gas jets for stabilization and photoelectric cells to maintain orientation. The spacecraft was controlled by radio command from Earth.

Luna 3 was launched into a highly elliptical Earth orbit that enabled the spacecraft to swing behind the Moon on October 6, with a minimum distance of about 6200 km (3853 mi). The dual-lens TV camera took 29 photographs over 40 min, imaging 70% of the unseen, but sunlit, far side. Images were recorded on photographic film, then developed, fixed, and dried on board before being scanned for transmission to Earth. The figure-eight flight path carried the spacecraft over the Moon's southern hemisphere, then diverted northwards above Earth–Moon plane and back towards Earth. Contact with the probe was lost on October 22.

TIROS 1 (Television and Infrared Observation Satellite 1, TIROS-I) USA

First operational weather satellite

SPECIFICATION:

MANUFACTURER: NASA
LAUNCH DATE: April 1, 1960
ORBIT: 692 x 740 km (430 x 460 mi), 48.4° inclination
LAUNCH SITE: Cape Canaveral, Florida
LAUNCHER: Thor-Able
LAUNCH MASS: 120 kg (270 lb)
DIMENSIONS: 1.1 x 0.6 m (3.5 x 1.8 ft)
PAYLOAD: 2 TV cameras

TIROS 1 began as a joint NASA/DOD project to develop a weather and reconnaissance satellite. Power was supplied by batteries charged by 9200 solar cells mounted on the sides and top of the spacecraft. Three pairs of solid-propellant rockets on the base plate controlled its spin.

The spacecraft was spin-stabilized and space-oriented (not Earth-oriented). As it rotated, the cameras scanned the clouds and surface. Each camera could take 16 pictures per orbit. The wide-angle camera imaged an area roughly 1207 km^2 (750 mi^2), the narrow angle an area about 129 km^2 (80 mi^2). When TIROS was in range of a ground station, the cameras could be commanded to take a picture every 10 or 30 sec.

The mission ended on June 15, 1960, after 78 days. TIROS 1 returned 22,952 cloud-cover photos. Nine more TIROS satellites were subsequently launched, carrying infrared radiometers as well as TV cameras. The last of these, TIROS 10, was launched on July 2, 1965, and shut down on July 3, 1967. Altogether, more than 500,000 cloud images were returned.

Corona (Discoverer, KH-1–KH-4B) USA

First US surveillance satellite

Corona was the code name for America's first space photo reconnaissance satellite program. Approved by President Eisenhower in February 1958, its objective was to take pictures of secret installations and military bases in the Soviet Bloc countries. The initial Corona launches were hidden as part a space technology program called Discoverer.

The first test launches took place in early 1959, and the first launch with a camera (Discoverer 4) was in June 1959. The first 13 missions did not return any images. The first success came on August 18, 1960, when a KH-1 obtained images of more than 4.27 million km² (1.65 million mi²) of Soviet territory on 915 m (3000 ft) of 70 mm film. The first Corona photo showed a Soviet air base on the shores of the Arctic Ocean. While in orbit, Corona took photographs with a constant rotating panoramic camera system and loaded the exposed photographic film into recovery vehicles. There were 145 Corona missions between 1959 and 1972. The last images were taken on May 31, 1972. A total of more than 800,000 images were taken on 640,000 m (2.1 million ft) of film stored in 39,000 cans.

SPECIFICATION: (KH-1)

MANUFACTURER: Lockheed Space Systems
LAUNCH DATE: February 28, 1959
ORBIT: 210 x 415 km (131 x 258 mi), 82.9° inclination (variable)
LAUNCH SITE: Vandenberg, California
LAUNCHER: Thor-Agena A
LAUNCH MASS: 779 kg (1 717 lb)
DIMENSIONS: 5.9 x 1.5 m (19.5 x 5 ft)
PAYLOAD: 1 or 2 convergent pan cameras; stellar terrain camera

Vostok 1 USSR

First human spaceflight

Launched into space by a modified R-7 ballistic missile, Vostok comprised a spherical reentry capsule, which housed the cosmonaut, a instrumentation section, and a rocket engine. There was no backup engine. The spacecraft was designed to operate automatically, with minimal involvement by the cosmonaut. Secret codes to unlock the controls were placed in an envelope for use in an emergency.

Vostok 1 completed one orbit of the Earth and the mission lasted for 108 min. The liquid-fuel engine fired for about 42 sec as the space-craft approached the coast of Angola. The instrumentation module initially failed to separate from the reentry module, so Gagarin experienced 10 min of buffeting before the wires burned through.

In the absence of a soft-landing system, Gagarin ejected at an altitude of 7 km (4.3 mi). He parachuted separately to the ground—though this was kept a secret for many years. The landing took place near the village of Smelovka, Saratov.

SPECIFICATION:
MANUFACTURER: OKB-1 (Korolev)
LAUNCH DATE: April 12, 1961
ORBIT: 181 x 327 km (112 x 203 mi), 64.95° inclination
LAUNCH SITE: Baikonur, Kazakhstan
LAUNCHER: Vostok (R-7)
LAUNCH MASS: 4730 kg (10,406 lb)
DIMENSIONS: 2.3 m (7.5 ft)
CREW: Yuri Gagarin

Mercury Redstone 3 (Freedom 7) USA

First US human spaceflight

SPECIFICATION (MR-3):

MANUFACTURER: McDonnell
LAUNCH DATE: May 5, 1961
ORBIT: suborbital, 187.5 km
(116.5 mi) altitude
LAUNCH SITE: Cape Canaveral,
Florida
LAUNCHER: Redstone
LAUNCH MASS: 1290 kg (2838 lb)
DIMENSIONS: 4 x 1.9 m (13.2 x
6.2 ft)
CREW: Alan Shepard

Mercury was America's first man-in-space program. The capsule was as small as possible to match the payload capability of the Redstone rocket, which was used for suborbital flights, and the Atlas, which was used for orbital missions.

The capsule comprised a cone-shaped crew compartment, topped by a cylindrical recovery compartment, antenna housing, and escape tower. Its outer skin was made of titanium and nickel alloy. The cabin atmosphere was 100% oxygen at a pressure of 0.36 kg/cm². (5.5 lb per in²). There were two independent thrusters systems for attitude control. Reentry was achieved by firing three solid propellant engines. During launch, Shepard experienced a peak g-load of 6.3. The Redstone's engine shut down on schedule after 142 sec, having accelerated the spacecraft to a speed of velocity of 8262 km/h (5133 mph). During about 5 min of weightlessness, Shepard experimented with controlling the spacecraft's attitude and observed the view through the periscope.

After the reentry engines fired, he experienced loads of up to 11 g. Following splashdown in the Atlantic, 487 km (302 mi) from Cape Canaveral, the capsule floated upright and Shepard was transported by helicopter to the carrier *Lake Champlain*, followed soon after by his capsule. The flight lasted 15 min 22 sec.

Telstar 1 USA

First operational communications satellite

SPECIFICATION:

MANUFACTURER: Bell Telephone Laboratories
LAUNCH DATE: July 10, 1962
ORBIT: 952 km x 5632 km (592 x 3500 mi), 44.8° inclination
LAUNCH SITE: Cape Canaveral, Florida
LAUNCHER: Delta
LAUNCH MASS: 77 kg (171 lb)
BODY DIMENSIONS: 0.9 m (2.9 ft)
PAYLOAD: Helical antenna; directional feed horns (transmission antennas)

Telstar was the first privately sponsored space launch and the first commercial communications satellite designed to transmit telephone and high-speed data. Owned by AT&T of the US, the project involved Bell Telephone Laboratories, NASA, the British Post Office, and the French National Post, Telegraph and Telecom Office. The main receiving station was Goonhilly Down in southern England.

The satellite was roughly spherical and spin-stabilized, its size limited by the dimensions of the Delta rocket fairing. The outer surface was covered by enough solar cells to produce 14 W of power. Rechargeable batteries were used to augment peak power requirements. On top was the main helical antenna. The 72 receive and 48 transmit antennas consisted of belts of small apertures around the satellite's "waist." It could handle 600 phone calls or one TV channel. Placed in an elliptical 2 h 37 min inclined orbit, it was only available to transmit signals across the Atlantic for 20 min during each orbit. The first live transatlantic TV pictures and the first telephone call transmitted via space were relayed on July 23, 1962.

Telstar was badly affected by increased radiation caused by high-altitude nuclear tests. It went out of service in early December 1962, but was restarted in early January 1963. The satellite was finally shut down on February 21, 1963. It remains in orbit around the Earth.

Mariner 2 USA

First successful Venus flyby and planet flyby

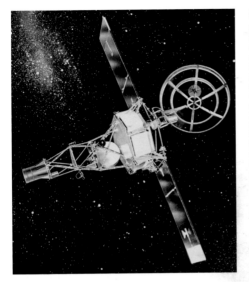

SPECIFICATION:

MANUFACTURER: Jet Propulsion Laboratory
LAUNCH DATE: August 27, 1962
ORBIT: Heliocentric
LAUNCH SITE: Cape Canaveral, Florida
LAUNCHER: Atlas-Agena B
LAUNCH MASS: 203.6 kg (448 lb)
DIMENSIONS: 3 x 5 m (9.9 x 16.5 ft) with antenna and solar panels deployed
PAYLOAD: Microwave radiometer; infrared radiometer; fluxgate magnetometer; cosmic dust detector; solar plasma spectrometer; energetic particle detectors

The Mariner series were the first US spacecraft sent to other planets. Mariner 1 and 2 were almost identical spacecraft developed to fly past Venus. The rocket carrying Mariner 1 went off-course during launch on July 22, 1962, and was destroyed by a range safety officer. A month later, Mariner 2 was launched on a 3½ month flight to Venus. On the way, it measured the solar wind and interplanetary dust. It also detected high-energy charged particles from the Sun and cosmic rays from outside the solar system.

The spacecraft flew by Venus at a range of 34,838 km (21,648 mi) on December 14, 1962. Mariner 2 scanned the planet for 42 min with infrared and microwave radiometers, revealing that Venus has cool clouds and an extremely hot surface. The spacecraft had a conical framework of magnesium and aluminum with two solar panels and a high-gain dish antenna. The data returned showed that the surface temperature on Venus was at least 425°C (800°F), with little variation between day and night. The planet's dense cloud layer extended from 56 to 80 km (34.8 to 49.7 mi) above the surface.

NASA maintained contact until January 3, 1963, when the spacecraft was 87.4 million km (54.3 million mi) from Earth, a new record for a deep space probe. The spacecraft remains in orbit around the Sun.

Syncom 1–3 USA

First geosynchronous communications satellite

SPECIFICATION:
(Syncom 2)

MANUFACTURER: Hughes Space
and Communications
LAUNCH DATE: July 26, 1963
ORBIT: 35,742 x 36,782 km
(22,210 x 22,856 mi), 33°
inclination (synchronous)
LAUNCH SITE: Cape Canaveral,
Florida
LAUNCHER: Thor-Delta
LAUNCH MASS: 35 kg (78 lb)
DIMENSIONS: 0.7 x 0.4 m (2.3 x
1.3 ft)
PAYLOAD: 2 1815 MHz
transmitters; 2 7363 MHz
receivers

The first Syncom was launched on February 14, 1963. At first, all proceeded smoothly. The ground team confirmed that the systems all functioned, but all communication was lost after the satellite's onboard apogee motor fired to inject Syncom into final orbit. Syncom 2's reliability was improved by reducing pressure in the nitrogen system and using a different motor. The satellite was a spin-stabilized cylinder coated with 3840 silicon solar cells.

On July 26, 1963, Syncom 2 successfully reached synchronous orbit over the Atlantic Ocean. With an orbital inclination of 33° it was not stationary over one point on the Earth's surface but moved in an elongated figure-eight pattern between 33°N and S of the equator. Its successes included the first live two-way call between heads of state by satellite relay.

By 1964, launch vehicle technology had advanced sufficiently for Syncom 3 to reach a geostationary orbit. Located over the Pacific Ocean, Syncom 3 was used to relay live TV coverage of the 1964 Olympic Games in Tokyo.

After the Department of Defense took over Syncom 2 and 3, they served as the primary communications link between Southeast Asia and the Western Pacific during the Vietnam conflict. The satellites were decommissioned and retired in April 1969.

Ranger 7 USA

First high-resolution images of the Moon (Ranger 6 pictured)

SPECIFICATION:

MANUFACTURER: Jet Propulsion Laboratory
LAUNCH DATE: July 28, 1964
ORBIT: Escape (impact)
LAUNCH SITE: Cape Canaveral, Florida
LAUNCHER: Atlas-Agena B
LAUNCH MASS: 366 kg (800 lb)
BODY DIMENSIONS: 3.6 x 1.5 m (11.8 x 5 ft)
PAYLOAD: 6 TV cameras

The Ranger series was the first US attempt to obtain close-up images of the Moon's surface. The spacecraft were designed to fly directly into the Moon and send images back until the moment of impact. The first five Rangers were failures. A modified version, known as Block III, was developed for Ranger 6, but no pictures were returned before lunar impact. Later investigations showed that sterilization procedures damaged components and contributed to failures.

Ranger 7 had a magnesium hexagonal framework to which two solar panels and a high-gain antenna were attached. The TV system consisted of six slow-scan vidicon cameras. The overlapping photographs taken by these cameras provided large-scale topographic information needed for the Surveyor and Apollo projects.

The camera system began transmitting pictures at 13:08 UT on July 31, 1964, 17 min 13 sec prior to impact. The last partial picture was taken about 0.3 sec before impact and achieved resolution of 0.5 m. In all, 43,816 black-and-white photographs were returned, showing a flat surface with numerous small craters. The spacecraft hit the Moon's Mare Nubium (10.35°S, 339.42°E) at 13:25:49 UT.

Mariner 4 USA

First successful Mars flyby

Two Mariner spacecraft were designed by NASA's Jet Propulsion Laboratory for the first close-up study of Mars, but Mariner 3 was destroyed after its launch shroud failed to separate. The identical Mariner 4 was launched 23 days later.

The spacecraft's main body was an octagonal magnesium frame with hydrazine thrusters for course correction. There were four solar panels and 12 nitrogen thrusters for attitude control.

The closest approach to Mars was 9846 km (6118 mi) on July 15, 1965. Forty min before, the TV scanning system began an automatic 25-min imaging sequence, returning 21 partially overlapping black-and-white pictures that covered a strip of land from about 37°N to 55°S—about 1% of the planet's surface. These pictures took four days to transmit to Earth. They showed a heavily cratered, ancient surface with no evidence of life. No magnetic field was detected.

Once past Mars, Mariner 4 orbited the Sun before returning to the vicinity of Earth in 1967. The spacecraft was used for operational and telemetry tests to improve knowledge of the technologies that would be needed for future interplanetary spacecraft. The final communication took place on December 21, 1967.

SPECIFICATION:

MANUFACTURER: Jet Propulsion Laboratory
LAUNCH DATE: November 28, 1964
ORBIT: Heliocentric
LAUNCH SITE: Cape Canaveral, Florida
LAUNCHER: Atlas-Agena D
LAUNCH MASS: 260.8 kg (574 lb)
BODY DIMENSIONS: 1.4 x 2.9 m (4.5 x 9.5 ft)
PAYLOAD: TV imaging system; cosmic dust detector; cosmic ray telescope; ionization chamber; magnetometer; trapped radiation detector; solar plasma probe

Voskhod USSR

First three-person spacecraft, first spacewalk (Voskhod 2 pictured)

SPECIFICATION
(Voskhod 2):

MANUFACTURER: OKB-1 (Korolev)
LAUNCH DATE: March 18, 1965
ORBIT: 167 x 475 km (103 x 295 mi), 64.8° inclination
LAUNCH SITE: Baikonur, Kazakhstan
LAUNCHER: Voskhod (R-7 with Luna second stage)
LAUNCH MASS: 5682 kg (12,526 lb)
DIMENSIONS: 5 x 2.4 m (16.4 x 8 ft)
CREW: Pavel Belyayev, Alexei Leonov

Voskhod (Sunrise) was a modified Vostok spacecraft, hurriedly adapted in advance of the US Gemini program. Only 6 months after the adaptation of two spacecraft began on April 13, 1964, Voskhod 1 went into orbit with a three-man crew. In order to make room for them, the Vostok ejection seat was removed and the crew were unable to wear pressure suits. The crew had no emergency escape system.

Shock-absorbing couches and a soft landing system allowed the crew to land inside the capsule. The orientation system used ion sensors, and a back-up solid propellant retrorocket was installed, making it safer to enter a higher orbit. Voskhod 2 was fitted with an inflatable cylindrical airlock.

The Voskhod 1 mission carried Boris Yegorov, Konstantin Feoktistov, and Vladimir Komarov. They completed 16 orbits in 24 h before landing northeast of Kustanai, Kazakhstan, on October 13, 1964. Voskhod 2, carrying Pavel Bela-yayev and Alexei Leonov, lasted 26 h, and included the world's first spacewalk. This 24 min extra-vehicular activity (EVA) almost ended in disaster when Leonov's suit ballooned, preventing him from reentering the spacecraft until he reduced its pressure. An unscheduled manual reentry meant landing in the Ural Mountains and a night in the forest before the crew were recovered.

Gemini 3 USA

First US multicrew flight

Project Gemini was the US program to develop techniques required for the Apollo Moon landings, such as rendezvous, docking, and EVA. After two unmanned test flights, the first manned mission was the first attempt to simulate rendezvous by changing orbit. The spacecraft, Gemini, was similar in shape to Mercury, but larger. At the rear was the large equipment section; in the center was the deorbit section, carrying four solid-fuel retro-rockets. In front was the pressurized cabin with a conical habitation section.

Gemini 3 was the first spacecraft to carry a computer. It weighed 22.7 kg (50 lb) and could make 7000 calculations per second. Each astronaut had a hatch, fitted with a porthole, large enough to permit an EVA. Later versions carried a rendezvous radar and fuel cells to make electricity.

Gemini 3 completed three orbits in 4 h 53 min. It was the first spacecraft to change orbit. This was achieved three times by using the 16 thrusters on the adapter section. Splashdown in the Atlantic was more than 80 km (50 mi) uprange from the predicted landing point.

SPECIFICATION:

MANUFACTURER: McDonnell Aircraft Corporation
LAUNCH DATE: March 23, 1965
ORBIT: 160 x 224 km (100 x 139 mi), 33° inclination
LAUNCH SITE: Cape Canaveral, Florida
LAUNCHER: Titan II
LAUNCH MASS: 3232 kg (7111 lb)
DIMENSIONS: 5.6 x 10 m max (18.4 x 32.8 ft)
CREW: Virgil "Gus" Grissom, John Young

Luna 9 USSR

First soft landing on the Moon, first pictures from lunar surface

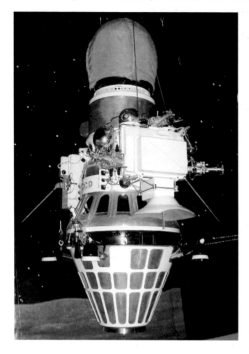

SPECIFICATION:

MANUFACTURER: Lavochkin
LAUNCH DATE: January 31, 1966
ORBIT: Heliocentric
LAUNCH SITE: Baikonur, Kazakhstan
LAUNCHER: R-7 (Vostok)
LAUNCH MASS: 1538 kg (3384 lb)
BODY DIMENSIONS: 2.7 x 1.2 m (8.9 x 3.9 ft); capsule diameter, 58 cm (1.9 ft)
PAYLOAD: TV imaging system; radiation detector

Luna 9 was the first spacecraft to achieve a lunar soft landing and then transmit photographs to Earth. It comprised three main sections. At the bottom was a liquid fuel retro-rocket. The central cylinder contained the communications and control units; strapped on either side were jettisonable compartments containing a radar altimeter, sets of nitrogen thrusters, and batteries. On top was a spheroidal container housing the landing capsule.

After making a direct approach to the Moon, the radar triggered the final descent sequence at an altitude of 75 km (47 mi). The two side compartments were jettisoned and the main engine ignited to slow the descent. A sensor detected the surface when the spacecraft was 5 m (16 ft) out and ordered the engine to shut down. The capsule was ejected while the main bus impacted at 22 km/h (14 mph) in the Mare Procellarum on February 3, 1966. The capsule, weighing 99 kg (218 lb), was hermetically sealed and contained radio equipment, a program timing device, heat control systems, scientific apparatus, power sources, and a TV system. Nine scans were completed during seven sessions totaling 8 h and 5 min.

Surveyor 1   USA

First US soft landing on the Moon

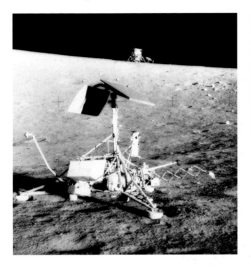

SPECIFICATION
(Surveyor 1):

MANUFACTURER: Hughes Aircraft
LAUNCH DATE: May 30, 1966
ORBIT: Heliocentric
LAUNCH SITE: Cape Canaveral,
Florida
LAUNCHER: Atlas-Centaur
LAUNCH MASS: 995 kg (2189 lb)
DIMENSIONS: 3 x 4.3 m (9.8 x
14 ft) with legs deployed
PAYLOAD: TV camera

Surveyor 1 was the first to spacecraft to make a truly soft landing on the Moon. Altogether, seven landers were built. Their main objectives were to send back detailed images from the lunar surface and to determine if the surface was safe for the Apollo manned landings.

Each Surveyor spacecraft was equipped with a TV camera. Surveyors 3 and 7 also carried a scoop for digging trenches and were used for soil mechanics tests. Surveyors 5, 6, and 7 had magnets attached to the footpads and an alpha scattering instrument for chemical analysis of soil. Surveyors 2 and 4 crashed on the Moon. The Surveyors had a triangular aluminum framework with a hinged leg at each corner and two panels—one solar array and a high gain antenna—above a central mast. The heaviest unit was the solid fuel retro-rocket, which was jettisoned after use. Surveyor 1 landed in the southwest region of the Mare Procellarum (2.5°S, 43.2°W) on June 2, 1966, 14 km (8.7 mi) from the planned target. At landing, the spacecraft weighed 294.3 kg (647 lb). During two communications sessions, having survived one lunar night, Surveyor 1 sent back 11,350 images, revealing a dusty plain with numerous boulders more than 1 m (3.3 ft) across and low mountains on the horizon. The mission was completed by July 13, but NASA maintained contact until January 7, 1967.

TACSAT I USA

Groundbreaking military communications satellite

SPECIFICATION:

MANUFACTURER: Hughes Aircraft Company
LAUNCH DATE: February 9, 1969
ORBIT: 107° W, GEO (later changed)
LAUNCH SITE: Cape Canaveral, Florida
LAUNCHER: Titan IIIC
LAUNCH MASS: 645 kg (1424 lb)
BODY DIMENSIONS: 2.8 x 7.6 m (9.3 x 25 ft)
PAYLOAD: 2 hard-limiting repeaters (UHF and SHF)

TACSAT I was the largest and most powerful communications satellite of its day, designed for the Department of Defense under the direction of the US Air Force Space and Missile Systems Organization. The experimental satellite was used by the US armed services to test the feasibility of using synchronous satellites for tactical communications with mobile military field units, aircraft, and ships.

The first satellite to be dual-spin-stabilized on its minor axis, using the Boeing-developed gyrostat, TACSAT's payload could be continuously pointed at Earth while the main body rotated. The cylindrical satellite was coated with solar cells. At the top was a then-unique array of three antenna systems. Since the satellite's strong signals were picked up by all types of terminals, including those with antennas as small as 0.3 m (1 ft) in diameter, remotely deployed field units were able to communicate with their headquarters and each other through the satellite. The high power of the satellite's transmissions permitted simultaneous access by many users. TACSAT was taken out of service on December 16, 1972.

Apollo 11 USA

First manned Moon landing

SPECIFICATION:

MANUFACTURER: North American (CSM—Combined Command Module and Service Module), Grumman (LM)
LAUNCH DATE: July 16, 1969
LUNAR ORBIT: 111 x 314 km (69 x 195 mi)
LAUNCH SITE: Kennedy Space Center, Florida
LAUNCHER: Saturn V
LAUNCH MASS: 43,961 kg (96,715 lb)
BODY DIMENSIONS: CM, 3.5 x 3.9 m (11.4 x 12.8 ft); SM, 7.5 x 3.9 m (24.6 x 12.8 ft); LM, 7 x 4.3 m (22.9 x 14.1 ft)
CREW: Neil Armstrong, Edwin "Buzz" Aldrin, Michael Collins

Apollo 11 followed two manned flights in Earth orbit and two manned flights in lunar orbit. The main living quarters en route to the Moon were in the conical Command Module (CM). One astronaut remained inside the CM while his two companions explored the lunar surface.

The Lunar Module (LM) comprised an octagonal descent stage, with a rocket engine and four legs. This was used as a launch pad for the ascent stage, which included the crew living quarters, LM controls, and an ascent engine.

After almost two orbits, the S-IVB engine reignited for 5 min 47 sec to boost the spacecraft to escape velocity—39,500 km/h (24,545 mph). Three days later, the CSM engine fired for 357 sec to brake into lunar orbit. On July 20, the LM *Eagle* separated from the CM *Columbia* on the far side of the Moon. *Eagle* landed on the Mare Tranquillitatis with about 20 sec of fuel left. The crew collected samples, planted the US flag, and deployed experiments during a 2 h 31 min EVA. The ascent module blasted off after 21 h 36 min on the surface. *Eagle* docked with *Columbia* and the CM splashed down in the Pacific on July 24 after jettisoning the LM and SM (Service Module).

Luna 16 USSR

First automated lunar soil sample return

SPECIFICATION:

MANUFACTURER: Lavochkin
LAUNCH DATE: September 12, 1970
INITIAL LUNAR ORBIT: 110 km (68.4 mi) circular, 70° inclination
LAUNCH SITE: Baikonur, Kazakhstan
LAUNCHER: Proton D-1-e
LAUNCH MASS: 5727 kg (12,600 lb)
BODY DIMENSIONS: 4 x 4 m (13.1 x 13.1 ft)
PAYLOAD: TV stereo imaging system; remote arm and drill for sample collection; radiation detector

Luna 16 was the first success out of six Soviet attempts to land an automated craft on the Moon and return a soil sample to Earth – though it was too late to beat the Apollo 11 and 12 missions. The descent stage was equipped with a TV camera, radiation and temperature monitors, communications equipment, and an extendable arm with a drill for collecting a lunar soil sample. The ascent stage included a liquid-fuel engine beneath a smaller cylinder that carried a sealed soil sample container inside a spherical reentry capsule.

Luna 16 was launched toward the Moon from an Earth parking orbit. It entered a circular 110 km (68.4 mi) lunar orbit on September 17, 1970. After several orbit maneuvers, the main engine was fired on September 20 to begin the descent. The landing thrusters cut off at 2 m (6.6 ft), followed by free fall to the surface. Luna 16 landed on the Mare Fecunditatis, on the lunar night side.

The drill was able to penetrate to a depth of 35 cm (1.1 ft). After 26 h 25 min on the lunar surface, the ascent stage lifted off from the Moon carrying 101 g (0.2 lb) of material. The reentry capsule returned directly to Earth and made a ballistic entry into the atmosphere on September 24. The capsule landed about 80 km (50 mi) southeast of the city of Dzhezkazgan in Kazakhstan.

Luna 17 (Lunokhod 1) USSR

First automated lunar rover

Luna 17 was a Luna sample return vehicle with the return section replaced by the Lunokhod 1 rover. It launched from an Earth parking orbit towards the Moon and entered lunar orbit on November 15, 1970. The spacecraft soft-landed on the Sea of Rains on November 17. The 756 kg (1663 lb) rover comprised a pressurized magnesium alloy compartment and was controlled by a team of five in the Crimea. Communication sessions lasted about 6 h per day. Four TV cameras on the sides of the vehicle provided side, down, and rear views when the rover was stationary. Two front-mounted TV cameras provided stereo images during the drives. The rover traveled at up to 2 km/h (1.3 mph). Operations officially ended on October 4, 1971 after 11 lunar days. Lunokhod had traveled 10.5 km (6.5 mi), transmitting over 20,000 TV pictures and 206 panoramas. It had also conducted more than 500 soil tests. The rover was left parked on the Moon so that its laser reflectors could continue to be used.

SPECIFICATION: (Lunokhod 1)

MANUFACTURER: Lavochkin
LAUNCH DATE: November 10, 1970
ORBIT: Escape
LAUNCH SITE: Baikonur, Kazakhstan
LAUNCHER: Proton D-1-e
LAUNCH MASS: 5700 kg (12,540 lb)
BODY DIMENSIONS: 1.4 x 2.2 m (4.4 x 7 ft)
PAYLOAD: 2 TV cameras; 4 high-resolution photometers; X-ray spectrometer; penetrometer; laser reflector; radiation detectors; X-ray telescope; odometer/speedometer

Salyut 1 ussr

First space station

SPECIFICATION:

MANUFACTURER: OKB-1 (Korolev)
LAUNCH DATE: April 19, 1971
ORBIT: 184 x 214 km (114 x 132 mi) variable, 51.6° inclination
LAUNCH SITE: Baikonur, Kazakhstan
LAUNCHER: Proton
LAUNCH MASS: 18,500 kg (40,700 lb)
BODY DIMENSIONS: 14.4 x 4.2 m max (47.2 x 13.6 ft)
PAYLOAD: Orion telescope; Oasis greenhouse
CREW: Georgi Dobrovolsky, Vladislav Volkov, Viktor Patsayev

Salyut (Salute) 1 was a "civilian" space station for Earth observation and microgravity research. The largest component was the unpressurized service module which housed the fuel, oxygen, and water tanks. The central section was the main work and living area. At the front was the docking equipment and airlock/transfer tunnel, as well as a telescope and communications equipment. Most of its major components were originally made for OKB-52's Almaz military space station program, with others borrowed from Soyuz.

A few hours after liftoff, the station's engine was used to circularize the orbit. Three days later, Soyuz 10 was sent to link up with the station. A problem with the Soyuz docking mechanism prevented a pressure-tight docking, so the three man crew were unable to enter Salyut.

A second crew arrived in Soyuz 11 on June 7. This time docking was successful. The crew began a three-week program of scientific and bio-medical research. On June 16, the crew reported a strong smell of burning and smoke in the station, but after an emergency retreat to the Soyuz the atmosphere cleared.

The crew returned to Earth in the Soyuz on June 29, but were found dead on landing. Without pressure suits, the men had suffocated due to an air leak and depressurization of the reentry capsule. Salyut 1 remained unoccupied and was deorbited on October 11, after some 2800 orbits.

Mariner 9 USA

First planet orbiter

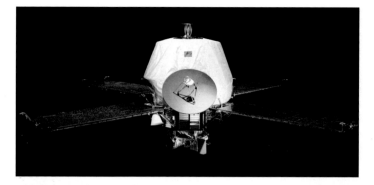

Mariner 8 and 9 were designed to be the first Mars orbiters. Mariner 8 was lost during launch on May 8, 1971, but the identical Mariner 9 was launched successfully three weeks later. The spacecraft was based on Mariner 6 and 7, with an octagonal magnesium framework and large propellant tanks for the retro-rocket.

After about a 5 month journey from Earth, Mariner 9 entered orbit around Mars on November 13, 1971. However, a dust storm was obscuring the entire planet. The storm lasted for a month, but then Mariner 9 was able to send back remarkable images of gigantic volcanoes, a huge canyon system, and dry river beds. Spacecraft tracking revealed a large equatorial bulge, and the first detection of water vapor was made over the south pole. Data were also returned on atmospheric composition, density, pressure, temperature, and surface composition. Mariner 9 also took the first close-up pictures of the two small Martian moons, Phobos and Deimos.

By the end of the mission, Mariner 9 had returned 7329 images covering the entire planet at 1–2 km resolution. The mission ended on October 27, 1972, when the supply of nitrogen for attitude control ran out.

SPECIFICATION:

MANUFACTURER: NASA–Jet Propulsion Laboratory
LAUNCH DATE: May 30, 1971
MARS ORBIT: 1394 x 17,144 km (866 x 10,653 mi), 64.34° inclination
LAUNCH SITE: Cape Canaveral, Florida
LAUNCHER: Atlas Centaur 23
LAUNCH MASS: 998 kg (2196 lb)
BODY DIMENSIONS: 2.3 x 6.9 m (7.5 x 22.6 ft) with solar panels deployed
PAYLOAD: TV system; ultraviolet spectrometer; infrared radiometer; infrared spectrometer

Pioneer 10 USA

First Jupiter flyby, first Saturn flyby, first spacecraft to leave the solar system

Pioneer 10 and 11 paved the way for the more advanced Voyagers. In 1972, Pioneer 10 was the fastest spacecraft ever to fly, leaving Earth at 51,670 km/h (32,107 mph). A gold anodized plaque bolted to the spacecraft's main frame marked its point of origin.

The first images of Jupiter were returned on November 6, 1973. Closest approach to the planet was on December 3, 1973, at a distance of 130,000 km (80,780 mi). More than 300 images were returned, many of them at a higher resolution than the best ground-based views. Other instruments studied the atmospheric circulation and composition, the radiation belts, the magnetic field, and the particle environment. Tracking enabled an accurate determination of the planet's mass.

The Jupiter flyby formally ended on January 2, 1974. Pioneer 10 crossed Neptune's orbit in May 1983, and for many years was the most remote man-made object in the solar system. The science mission continued until March 31, 1997. The last signal was received on January 23, 2003. The spacecraft is heading for the star Aldebaran in the constellation of Taurus. It will take Pioneer over two million years to reach it.

SPECIFICATION:

MANUFACTURER: TRW Systems Group, Inc.
LAUNCH DATES: March 3, 1972 (Pioneer 10); April 5, 1973 (Pioneer 11)
ORBIT: escape
LAUNCH SITE: Cape Canaveral, Florida
LAUNCHER: Atlas/Centaur/ TE364-4
LAUNCH MASS: 270 kg (594 lb)
BODY DIMENSIONS: 2.9 x 2.7 m (9.5 x 8.9 ft)
PAYLOAD: Helium vector magnetometer (France); plasma analyzer; charged particle instrument; cosmic ray telescope; Geiger Tube telescope; trapped radiation detector; meteoroid detector; Asteroid-Meteoroid Experiment; ultraviolet photometer; imaging photopolarimeter; infrared radiometer; flux gate magnetometer (Pioneer 11 only)

Skylab USA

First US space station

SPECIFICATION:
(Skylab 1)

MANUFACTURER: McDonnell Douglas Astronautics Co.
LAUNCH DATE: May 14, 1973
ORBIT: 431 x 434 km (268.1 x 269.5 mi), 50° inclination
LAUNCH SITE: Cape Kennedy, Florida
LAUNCHER: Saturn V (2 stage)
LAUNCH MASS: 74,783 kg (164,868 lb)
BODY DIMENSIONS: 25.3 x 6.7 m (83 x 22 ft) max.
PAYLOAD: 6 UV/X-ray telescopes (Apollo Telescope Mount); multispectral cameras; earth terrain camera; infrared spectrometer; multispectral scanner; microwave radiometer/scatterometer and altimeter; L-band radiometer; materials-processing facility; multipurpose electric furnace
CREWS: Pete Conrad, Paul Weitz, Joe Kerwin (Skylab 2); Alan Bean, Jack Lousma, Owen Garriott (Skylab 3); Jerry Carr, William Pogue, Edward Gibson (Skylab 4)

Skylab, based on the S-IVB third stage of the Saturn V rocket, was America's first space station and the largest spacecraft ever placed in Earth orbit. The first crew arrived 11 days later. Three crews visited the station, with missions lasting 28, 59, and 84 days (a record not broken by a US astronaut until the Shuttle–Mir program over 20 years later). They performed UV astronomy experiments and X-ray studies of the Sun, remote sensing of the Earth, and biological and medical studies.

The largest section of Skylab was the 295 m³ (10,246 ft³) workshop, which housed crew quarters plus lab facilities. In front were an airlock module and the docking adapter, to be used by an Apollo CSM spacecraft. The airlock was the environmental, electrical and communications control center.

Skylab was launched with a fixed amount of supplies—fuel, water, air, food, clothing, etc. After the final crew left on February 8, 1974, the station was put into a stable attitude and shut down. It was expected to remain in orbit for 8–10 years, but increased atmospheric drag caused it to reenter the atmosphere on July 11, 1979. The debris dispersion area covered the southeast Indian Ocean and a sparsely populated area of Western Australia.

Mariner 10 USA

First Mercury flybys

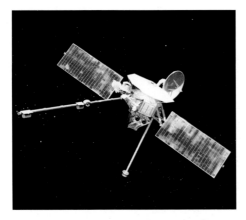

SPECIFICATION:

MANUFACTURER: Jet Propulsion Laboratory
LAUNCH DATE: November 3, 1973
ORBIT: Heliocentric
LAUNCH SITE: Cape Canaveral, Florida
LAUNCHER: Atlas-Centaur 34
LAUNCH MASS: 502.9 kg (1106 lb)
BODY DIMENSIONS: 1.4 x 0.5 m (4.6 x 1.5 ft)
PAYLOAD: TV imaging system; ultraviolet spectrometer; infrared radiometer; ultraviolet occultation spectrometer; solar plasma analyzer; charged particle telescope; 2 magnetometers

Mariner 10 was the first (and, so far, only) spacecraft to explore Mercury. This was achieved through the first use of a gravity-assist maneuver, the spacecraft passing close to Venus in order to bend its flight path and change its perihelion to the same solar distance as Mercury. The spacecraft then entered an orbit that repeatedly brought it back to Mercury. The three-axis-stabilized spacecraft was based on the Mariner Mars craft. Three months after leaving Earth, Mariner 10 passed Venus on February 5, 1974, at a distance of 5768 km (3584 mi). Altogether, 4165 photos of Venus were returned.

The first Mercury encounter took place on March 29, 1974, at a range of 703 km (437 mi). Looping once around the Sun while Mercury completed two orbits, Mariner 10 made another flyby on September 21, 1974, at a distance of 48,069 km (29,870 mi). The third and final encounter took place on March 16, 1975, at a range of 327 km (203 mi). Unfortunately, the geometry of Mariner 10's orbit meant that the same side of Mercury was sunlit each time, so only about 45% of the surface was seen. The images returned by Mariner 10 showed a heavily cratered world very much like the Moon. The most striking feature was a partially sunlit impact basin 1300 km (808 mi) across.

Venera 9 USSR

First images from surface of Venus, first Venus orbiter

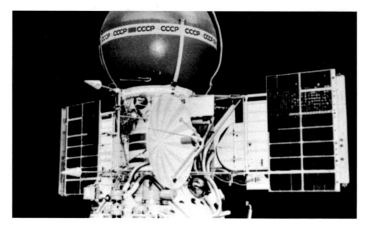

Venera 9 was the first of the second-generation Venus craft designed by the USSR. It comprised a conical base containing the electronics and the retro-rocket, with scientific instruments around its exterior, topped by the 2.4 m (7.9 ft) spherical descent module.

The orbiter entered a highly elliptical Venus orbit on October 20, 1975, after about 4.5 months. Two days later the descent sphere was released, entering the atmosphere at 10.7 km/s (6.6 mi/s) at an angle of 20.5°. Parachutes were then used, with the final descent in free fall. Touchdown was at about 7.5 m/s (26.6 ft/s). The lander recorded light levels comparable to those on a cloudy summer day on Earth, so its floodlights were not needed. Surface pressure was about 90 Earth atmospheres, and surface temperature 485°C (905°F). Precooling prior to entry and a system of circulating fluid to distribute the heat load enabled the craft to survive for 53 min. The first black-and-white images of the surface showed numerous rocks 30–40 cm (about 1 ft) across. The end of the orbiter mission was announced on March 22, 1976.

SPECIFICATION:

MANUFACTURER: Lavochkin
LAUNCH DATE: June 8, 1975
VENUS ORBIT: 1500 x 111,700 km (932 x 69,410 mi), 34.1° inclination
LAUNCH SITE: Baikonur, Kazakhstan
LAUNCHER: Proton D-1-e
LAUNCH MASS: 4936 kg (10,860 lb)
DIMENSIONS: 2.8 x 6.7 m (9.2 x 22 ft) with solar panels deployed
PAYLOAD, ORBITER: TV imaging system; UV imaging spectrometer; infrared radiometer; photopolarimeter; magnetometer; ion/electron detectors; optical spectrometer
PAYLOAD, LANDER: Panoramic imaging system; mass spectrometer; temperature and pressure sensors; anemometer; nephelometer; photometers; gamma-ray spectrometer; radiation densitometer; accelerometers

Viking USA

First US Mars landings

The two Vikings followed on from NASA's successful Mariner Mars missions. Each mission consisted of an orbiter and a lander. The orbiter was based on Mariner 9 but carried a much larger liquid-fuelled retrocket.

Viking 1 arrived at Mars on June 19, 1976. The lander separated from the orbiter and touched down at Chryse Planitia (22.48°N, 49.97°W) on July 20. Viking 2 touched down at Utopia Planitia (47.97°N, 225.74°W) on September 3, 1976. The primary mission objectives were to obtain high-resolution images of the Martian surface, characterize the structure and composition of the atmosphere and surface, and search for evidence of life.

The Viking landers sent back over 4500 images. The Viking 2 lander mission ended on April 11, 1980, and the Viking 1 lander shut down on November 13, 1982. The biology experiments found no clear evidence of living microorganisms. The orbiters provided more than 50,000 images, mapping 97% of the planet. The Viking 2 orbiter lasted until July 25, 1978, and the Viking 1 mission ended on August 7, 1980, after over 1400 orbits.

SPECIFICATION:

MANUFACTURER: Martin Marietta
LAUNCH DATES: August 20, 1975 (Viking 1); September 9, 1975 (Viking 2)
INITIAL MARS ORBIT: 1510 x 32,800 km (938 x 20,380 mi), 37.9° inclination
LAUNCH SITE: Cape Canaveral, Florida
LAUNCHER: Titan IIIE-Centaur
LAUNCH MASS: 3527 kg (7760 lb)
BODY DIMENSIONS: 3.3 x 2.5 m (10.8 x 8.2 ft)
PAYLOAD, ORBITER: 2 vidicon cameras; infrared spectrometer; infrared radiometer
PAYLOAD, AEROSHELL: Retarding potential analyzer; mass spectrometer; pressure, temperature, and acceleration sensors
PAYLOAD, LANDER: 2 TV cameras; biology lab (gas exchange, labeled release, and pyrolytic release); gas chromatograph mass spectrometer; X-ray fluorescence spectrometer; pressure, temperature, and wind velocity sensors; three-axis seismometer; sampler arm; magnets

Voyager USA

First Uranus and Neptune flybys, first exploration of heliopause

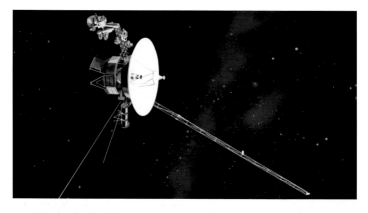

The two identical Voyager spacecraft were built to visit Jupiter and Saturn with a prime mission lasting five years. However, Voyager 2 was able to take advantage of a rare alignment of the outer planets that allowed gravity assists leading to flybys of Uranus and Neptune.

Voyager 2 launched first but was overtaken by Voyager 1 which flew past Jupiter on March 5, 1979, at a distance of 206,700 km (128,400 mi) from the cloud tops. Voyager 2's flyby was on July 9, 1979, at a range of 570,000 km (350,000 mi). The Voyager 1 Saturn flyby was on November 12, 1980, with close-up observations of Titan. Voyager 2 passed 41,000 km (26,000 mi) from Saturn on August 25, 1981.

Voyager 1's Titan encounter sent it on a path 35.5°N of the ecliptic, so no more planets could be visited. Voyager 2 flew past Uranus on January 24, 1986. Finally, it flew over Neptune's north pole at a distance of 5000 km (3000 mi) on August 25, 1989. The scan platform instruments have been turned off, but both spacecraft continue to send back data on interplanetary space and to search for the outer edge of the solar system.

SPECIFICATION:

MANUFACTURER: NASA–Jet Propulsion Laboratory
LAUNCH DATES: August 20, 1977 (Voyager 2); September 5, 1977 (Voyager 1)
ORBIT: Escape
LAUNCH SITE: Cape Canaveral, Florida
LAUNCHER: Titan IIIE – Centaur
LAUNCH MASS: 825 kg (1815 lb)
BODY DIMENSIONS: 0.5 x 1.8 m (1.5 x 5.8 ft) without high-gain antenna
PAYLOAD: Imaging system; infrared spectrometer; ultraviolet spectrometer; photopolarimeter; planetary radio astronomy; magnetometers; low energy charged particle investigation; plasma particles investigation; cosmic ray telescope; plasma wave investigation

Pioneer Venus 1 (Pioneer Venus Orbiter) USA

First US Venus orbiter, first surface radar mapper

SPECIFICATION:

MANUFACTURER: NASA Ames
LAUNCH DATE: 20 May 1978
VENUS ORBIT: 150 x 66,889 km
(93 x 41,564 mi), 105° inclination
LAUNCH SITE: Cape Canaveral,
Florida
LAUNCHER: Atlas-Centaur
LAUNCH MASS: 517 kg (1137 lb)
BODY DIMENSIONS: 2.5 x 1.2 m
(8.3 x 4 ft)
PAYLOAD: Cloud photopolarimeter; surface radar mapper;
infrared radiometer; airglow
ultraviolet spectrometer; neutral
mass spectrometer; solar wind
plasma analyzer; magnetometer;
electric field detector; electron
temperature probe; ion mass
spectrometer; charged particle
retarding potential analyzer;
gamma-ray burst detector

Pioneer Venus 1 (PV1) was part of a two-pronged exploration of Venus. Its mission was to study the planet's thick clouds and upper atmosphere and to make the first map of the hidden surface. Six months after launch, on December 4, 1978, the orbiter was injected into a highly elliptical orbit around Venus. Circling the planet every 23 h 11 min, it passed within 150 km (93 mi) of the surface. Radar mapping took place around the close approaches. Although the radar instrument malfunctioned between December 18, 1978, and January 20, 1979, it mapped most of the surface between 73°N and 63°S at a resolution of 75 km (47 mi). The map revealed broad plains, two large "continents," and many volcanic features.

PV1 also recorded global observations of the clouds and atmosphere and made studies of the solar wind–ionosphere interaction. For the first 19 months, the periapsis remained at about 150 km (93 mi). As propellant began to run low, the maneuvers were discontinued, and the low point rose to about 2300 km (1430 mi). From 1986, the orbit lowered again, enabling further measurements within the ionosphere. PV1 also observed several comets with its ultraviolet spectrometer, including Halley's Comet.

On October 8, 1992, with its fuel supply exhausted, the spacecraft burnt up in the planet's atmosphere.

Pioneer Venus 2 (Pioneer Venus Multiprobe) USA

First US Venus atmospheric probes

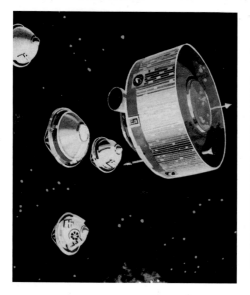

SPECIFICATION:

MANUFACTURER: NASA Ames
LAUNCH DATE: August 8, 1978
ORBIT: Heliocentric (impact)
LAUNCH SITE: Cape Canaveral, Florida
LAUNCHER: Atlas-Centaur
LAUNCH MASS: 875 kg (1925 lb); bus 290 kg (638 lb); large probe 315 kg (693 lb); small probes 90 kg (204.6 lb) each
BODY DIMENSIONS: 2.5 x 1.2 m (8.3 x 4 ft)
PAYLOAD: Neutral mass spectrometer (bus, large probe); ion mass spectrometer (bus); gas chromatograph (large probe); solar flux radiometer (large probe); infrared radiometer (large probe); cloud particle size spectrometer (large probe); nephelometer (large probe, small probes); temperature, pressure, and acceleration sensors (large probe, small probes); net flux radiometer (small probes)

Pioneer Venus 2 consisted of a bus that carried four atmospheric probes. Their mission was to make a ballistic entry into the planet's dense, hot atmosphere and send back data to Earth during their descent. After a three-month journey to Venus, the large probe was released from the bus on November 16, 1978, followed by the small probes on November 20. All four probes entered the planet's atmosphere on December 9.

The drum-shaped bus was similar to the Pioneer Venus Orbiter but without a high-gain antenna and orbit insertion motor. Although it had no heat shield, its instruments were able to return data before it burnt up in the upper atmosphere.

The large probe was about 1.5 m (4.9 ft) in diameter. It carried seven science experiments. The three identical small probes were 0.8 m (2.6 ft) in diameter. The radio signals from all four probes were used to characterize the winds, turbulence, and other characteristics of the atmosphere.

The small probes were each targeted at different parts of the planet. The North Probe entered the atmosphere at 59.3°N, 4.8°E on the day side. The Night Probe entered on the night side at 28.7°S, 56.7°E. The Day Probe entered at 31.3°S, 317°E, well into the day side, and was the only one to send radio signals back after impact, lasting for over an hour.

ISEE-3/ICE (International Sun-Earth Explorer 3/ International Cometary Explorer, Explorer 59) USA

First comet encounter, first craft in halo orbit

SPECIFICATION:

MANUFACTURER: Fairchild
LAUNCH DATE: August 12, 1978
ORBIT: L1 halo, then heliocentric
LAUNCH SITE: Cape Canaveral, Florida
LAUNCHER: Delta 2914
LAUNCH MASS: 478 kg (1052 lb)
BODY DIMENSIONS: 1.7 x 1.6 m (5.6 x 5.3 ft)
PAYLOAD: Solar Wind Plasma Experiment; Magnetometer; X-Ray and Low Energy Cosmic Ray Experiment; Interplanetary and Solar Electrons Experiment; Plasma Wave Experiment; Plasma Composition Experiment; Radio Wave Experiment; energetic particle anisotropy spectrometer; X-ray and gamma-ray bursts; medium-energy cosmic rays and electrons; high-energy cosmic rays; cosmic ray energy spectrum; cosmic ray isotope spectrometer

This NASA spacecraft was launched into a halo orbit around the L1 Sun-Earth libration point, 1.5 million km on the sunward side of the Earth. Its mission was to provide data on the solar wind.

On June 10, 1982, having completed its original mission, it was diverted towards Earth and began a series of lunar flybys. A final close lunar flyby, 119.4 km (74.2 mi) above the Moon's surface, on December 22, 1983, ejected the spacecraft into a heliocentric orbit that would enable it to intercept Comet Giacobini-Zinner. ISEE-3 was then renamed the International Cometary Explorer (ICE).

ICE made the first ever passage through a comet's tail on September 11, 1985, sending back data on particles, fields, and waves. It passed 7860 km (4884 mi) behind the comet's nucleus at a closing speed of 75,300 km/h (46,790 mph). Having survived the passage through the coma, ICE went on to observe Halley's Comet in March 1986, becoming the first spacecraft to directly investigate two comets. Operations ended May 5, 1997, but ICE remains in a 355-day heliocentric orbit that will bring it back to the vicinity of the Earth in August 2014.

IRAS (Infrared Astronomy Satellite) NETHERLANDS/USA/UK

First all-sky infrared observatory

IRAS was designed to survey the whole sky at infrared wavelengths. The satellite bus was built by the Dutch; the US provided the telescope, detectors, and launch while the UK provided the satellite control facilities.

The top section was built around a large dewar (vacuum vessel) filled with liquid helium. The outer wall of the dewar was insulated, and a sunshade prevented sunlight from entering the telescope. The lower section included the control systems and communications.

IRAS flew over the poles every 90 min and scanned a 0.5° wide strip of sky that half overlapped with its previous scan. It eventually scanned the entire sky three times. The mission ended when the supply of liquid helium ran out on November 21, 1983, after almost 10 months of operation. IRAS detected about 350,000 infrared sources, increasing the number of cataloged astronomical sources by about 70%. It also discovered six new comets and revealed for the first time the core of the Milky Way galaxy.

SPECIFICATION:

MANUFACTURER: Netherlands Institute for Aeronautics and Space
LAUNCH DATE: January 26, 1983
ORBIT: 889 x 903 km (552 x 561 mi), 99° inclination (sun-synchronous)
LAUNCH SITE: Vandenberg, California
LAUNCHER: Delta 3910
LAUNCH MASS: 1075.9 kg (2367 lb)
BODY DIMENSIONS: 3.6 x 3.2 m (11.8 x 10.6 ft)
PAYLOAD: Infrared camera (ISOCAM); short wavelength spectrometer (SWS); long wavelength spectrometer (LWS); infrared photometer and polarimeter (ISOPHOT)

Spacelab 1 (STS-9/STS-41A) EUROPE/USA

First European manned space facility

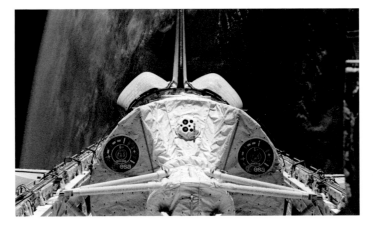

In 1973, the European Space Research Organiz-ation (later ESA) agreed with NASA to provide Spacelab, a manned science laboratory. The modular lab, comprising one long module and a single pallet, was designed and built by ESA under West German leadership.

The first Spacelab flew in the payload bay of the Shuttle *Columbia* on mission STS-9/STS-41A, November 28–December 8, 1983. A long tunnel connected the module to the Shuttle mid-deck.

STS-9 carried six people into orbit on a single vehicle. Among the crew were Byron Lichtenberg and Ulf Merbold, the first payload specialists to fly on the Shuttle. Merbold, from West Germany, was also the first foreign citizen to participate in a Shuttle flight. The crew was divided into two teams, each scheduled to work 12 h shifts but sometimes working up to 18 h at a time. The first direct voice communications were set up via the TDRS-1 satellite.

Seventy scientific experiments were carried out in many different fields. They were provided by 11 European nations, the US, Canada, and Japan.

SPECIFICATION:

MANUFACTURER: MBB-ERNO
LAUNCH DATE: November 28, 1983
ORBIT: 241 x 254 km (150 x 158 mi), 57° inclination
LAUNCH SITE: Kennedy Space Center, Florida
LAUNCHER: Space Shuttle (*Columbia*)
LAUNCH MASS: Pressure module, 8145 kg (17,919 lb); pallet, 3386 kg (7449 lb)
DIMENSIONS: Pressure module, 4.1 x 7 m (13.5 x 23 ft); pallet, 3 x 4 m (9.8 x 13.1 ft)
PAYLOAD: 6 astronomy and solar physics experiments; 6 space plasma physics experiments; 6 atmospheric and Earth observations; 16 life sciences experiments; 36 materials science experiments
CREW: John Young, Brewster Shaw, Owen Garriott, Robert A. Parker, Ulf Merbold, Byron Lichtenberg

Vega USSR

First close comet flyby, first planetary balloons

SPECIFICATION:

MANUFACTURER: Lavochkin
LAUNCH DATES: December 15,
1984 (Vega 1); December 21,
1984 (Vega 2)
ORBIT: Heliocentric
LAUNCH SITE: Baikonur,
Kazakhstan
LAUNCHER: Proton D-1-e
LAUNCH MASS: 4920 kg (10,824 lb)
BODY DIMENSIONS: 2.5 x 3 m
(8.2 x 9.8 ft)
PAYLOAD, LANDER: Mass
spectrometer; gas chromatograph; hygrometer; gamma-ray
spectrometer; ultraviolet
spectrometer; X-ray fluorescence
spectrometer and drill;
nephelometer/scatterometer;
temperature and pressure
sensors; aerosol analyzer
PAYLOAD, BUS: Television system
(TVS); infrared spectrometer
(IKS); three-channel (UV, visible,
infrared) spectrometer (TKS);
dust mass spectrometer (PUMA);
dust particle counters (SP);
neutral gas mass spectrometer
(ING); plasma energy analyzer
(PLASMAG); energetic particle
analyzer (TUNDE-M); magnetometer (MISCHA); wave and
plasma analysers (APV-N);
energetic particles (MSU-TASPD)
PAYLOAD, BALLOON: Temperature
and pressure sensors; vertical
wind anemometer; nephelo-
meter; light level sensors

The Soviet missions to explore Halley's Comet
were called "Vega," from "Venera-Gallei" (rather
than Venera-Halley) because there is no "H" in
the Russian alphabet. The identical spacecraft
were the last to be based on the second
generation Venera vehicles.

The spacecraft were launched six days apart in
December 1984. Their first task was to deliver a
lander and a Teflon-coated plastic balloon as they
flew past Venus. The Vega 1 lander was released
on June 9, 1985, followed by the second on June
15. On the way down, a balloon separated from
the 1500 kg descent module and inflated with
helium. Both balloons survived for about 46.5 h.

Retargeted by the Venus gravity assist, the
Vega 1 bus encountered Halley's Comet on March
6, 1986; Vega 2 did so three days later at a flyby
velocity of 77.7 km/s. Uncertainty in the position
of the comet nucleus and concern over dust
damage led to flyby distances of 39,000 km
(24,235 mi) for Vega 1 and 8030 km (4990 mi) for
Vega 2. Both spacecraft obtained images of the
nucleus and acted as pathfinders for ESA's Giotto
mission.

Giotto EUROPE

First ESA deep space mission, close flybys of two comets

SPECIFICATION:

MANUFACTURER: British Aerospace
LAUNCH DATE: July 2, 1985
ORBIT: Heliocentric
LAUNCH SITE: Kourou, French Guiana
LAUNCHER: Ariane 1
LAUNCH MASS: 960 kg (2112 lb)
DIMENSIONS: 1.9 x 2.9 m (6.1 x 9.3 ft)
PAYLOAD: Halley multicolor Camera; neutral mass spectrometer; ion mass spectrometer; dust mass spectrometer; dust impact detector; plasma analysis (2); energetic particle analyzer; magnetometer; optical probe

Giotto was named after Giotto di Bondone who portrayed Halley's Comet as the Star of Bethlehem in his fourteenth-century fresco *Adoration of the Magi*. It was originally proposed as part of a joint US–European mission to Halley's Comet, but NASA later withdrew.

The cylindrical spacecraft was based on the GEOS Earth-orbiting research satellites. The most significant addition was an aluminum and Kevlar bumper shield to protect it from high-speed dust particles during the comet encounter.

Giotto was part of an international program to investigate Halley. The first comet data arrived on March 12, 1986, when the instruments detected hydrogen ions 7.8 million km from the nucleus; 22 hours later, Giotto crossed the bow shock, where the solar wind slowed to subsonic speed, and entered the coma; 7.6 seconds before closest approach, it was sent spinning by a dust particle. Giotto lost contact with the Earth for 32 minutes, and its camera was irreparably damaged. But by then Giotto had passed 596 km (370 mi) from the nucleus, obtaining the closest pictures ever taken of a comet.

Giotto was redirected to Comet Grigg-Skjellerup. It was placed in hibernation on July 23, 1992, and the spacecraft has since been inactive.

Mir USSR

Third generation Soviet space station

SPECIFICATION:

MANUFACTURER: RSC Energia
LAUNCH DATES: February 20, 1986
(base module); March 31, 1987
(Kvant-1); November 26, 1989
(Kvant-2); May 31, 1990 (Kristall);
May 20, 1995 (Spektr); April 23,
1996 (Priroda)
ORBIT: 385 x 393 km (239 x
244 mi), 51.6° inclination
LAUNCH SITE: Baikonur,
Kazakhstan
LAUNCHER: Proton
MASS: 137,160 kg (135 tons)
BODY DIMENSIONS: 13.1 x 29.7 m
(43.1 x 97.5 ft)
CREWS: 104 different crew from
12 different nations (125 visitors
in total); 17 visiting expeditions
and 28 long term crews

Mir (Peace or World) was the successor to the Salyut series of Soviet/Russian space stations. The station's modular design resulted from the payload limitations of the Proton rocket. It eventually comprised seven modules, including a docking module for use by the Space Shuttle. All of the Soviet modules were delivered by Proton and docked automatically.

The three-section base module contained the living quarters. At the front was a separate airlock/five-port docking node. At the rear was the service section with thrusters, main engine, tanks for consumables, and communications.

On June 29, 1995, US Space Shuttles began docking with Mir. Before this first docking, the Mir-19 crew used the Lyappa manipulator arm to relocate Kristall module, providing clearance for *Atlantis*. In November 1995, a Russian-built docking module arrived on Shuttle mission STS-74 and was attached to Kristall. Finally, the Priroda remote sensing module was added opposite Kristall.

Over its lifetime, 31 crewed spacecraft docked with Mir, including nine Shuttles, as well as 64 uncrewed cargo ships. In 1997 the station suffered a serious fire and a collision with a Progress ship that caused partial depressurization. It was deorbited over the Pacific on March 23, 2001.

Buran USSR

First automated flight of a reusable shuttle

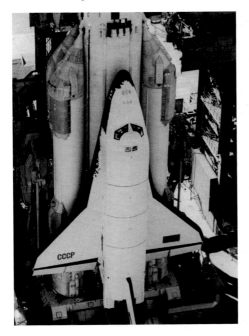

SPECIFICATION:

MANUFACTURER: NPO Molniya
LAUNCH DATE: November 15, 1988
ORBIT: 252 x 256 km, 51.6° inclination
LAUNCH SITE: Baikonur, Kazakhstan
LAUNCHER: Energia
LAUNCH MASS: 79,400 kg (175,000 lb)
LENGTH: 36.7 m (119.4 ft)
WINGSPAN: 23.9 m (78.4 ft)
MAX. HEIGHT (WITH WHEELS DEPLOYED): 16.4 m (53.8 ft)

The Buran (Snowstorm) orbiter was the Soviet response to the US Space Shuttle. It was the first (and, so far, only) reusable Russian manned space vehicle. The forward cabin could house a crew of six but the orbiter's only flight was unmanned. It was then abandoned because of cost.

Unlike the Shuttle, Buran carried no rocket engines. It was launched on the side of the Energia booster, which provided all of the launch thrust. The orbiter returned to Earth like a glider, landing on a runway. Thermal protection for the descent required almost 40,000 custom-made heat tiles.

Buran completed two orbits of the Earth. The Energia core stage shut down eight minutes after launch and the orbiter separated at 160 km (99.4 mi) altitude. About 2 min later, the orbiter's maneuvering engines fired for 67 sec to boost the altitude to about 250 km (155 mi). Another maneuver over the Pacific circularized the orbit. Retrofire occurred during the second pass over the Pacific, and the mission ended with an automated landing at Baikonur. The only flown Buran orbiter was destroyed when the Energia hangar collapsed in May 2002.

Hipparcos (High Precision Parallax Collecting Satellite) ESA

First astrometry mission

SPECIFICATION:

MANUFACTURER: Matra Marconi Space
LAUNCH DATE: August 8, 1989
ORBIT: 526 x 35,896 km (327 x 22,305 mi), 6.9° inclination
LAUNCH SITE: Kourou, French Guiana
LAUNCHER: Ariane 44LP
LAUNCH MASS: 1140 kg (2508 lb)
BODY DIMENSIONS: 3 x 1.8 m (9.8 x 5.9 ft)
PAYLOAD: 0.3 m (1 ft) Schmidt telescope; two star mappers

Hipparcos was the first space mission for astrometry—dedicated to the precise measurement of the positions, distances, and proper motions of the stars.

The satellite was to operate in GEO, but it was stranded in a highly elliptical transfer orbit when the solid-fuel boost engine failed to fire. The satellite had to endure repeated passes through the Van Allen radiation belts and periods in shadow. Despite these problems, Hipparcos outlived its 30-month design life before communications were terminated in August 1993.

Eventually, Hipparcos pinpointed more than 100,000 stars, 200 times more accurately than ever before. The spacecraft turned on its axis in just over 2 h, while slowly changing the direction of the axis of rotation so that it could repeatedly scan the entire sky. It measured angles between widely separated stars and recorded their brightness, which often varied over time. Each star selected for study was visited about 100 times over four years. The end product was two catalogs—the Hipparcos Catalog, charted with the highest precision, and the Tycho Catalog, more than 2.5 million stars mapped with slightly lower accuracy by an auxiliary star mapper.

Cosmic Background Explorer (COBE) USA

First detailed study of cosmic microwave background

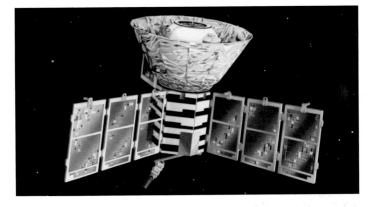

COBE was developed to provide a highly accurate measurement of the diffuse radiation from the early universe. It carried three instruments—to search for cosmic infrared background radiation, to map variations in the cosmic microwave radiation, and to compare the spectrum of the background radiation with a precise blackbody. Each yielded a major cosmological discovery.

Intended for launch on a Shuttle, it had to be downsized after the *Challenger* disaster to fit on a Delta rocket.

The experiment module on top contained the instruments and a dewar filled with 650 l (1.1 qt) of liquid helium at a temperature of 1.6 K (just above absolute zero), with a conical, deployable sun-shade. The satellite was turned at 94° from the Sun and away from Earth. The instruments performed a complete scan of the celestial sphere every six months.

After the helium ran out on September 21, 1990, operations continued with the onboard differential microwave radiometer until December 23, 1993. COBE was then used as an engineering training and test satellite by NASA's Wallops Flight Facility.

SPECIFICATION:

MANUFACTURER: NASA–Goddard
LAUNCH DATE: November 18, 1989
ORBIT: 900 km (559 mi), 99° inclination (sun-synchronous)
LAUNCH SITE: Vandenberg, California
LAUNCHER: Delta 5920
LAUNCH MASS: 2270 kg (4994 lb)
BODY DIMENSIONS: 2.5 x 4.9 m (8.2 x 16 ft)
PAYLOAD: Diffuse infrared background experiment (DIRBE); differential microwave radiometer (DMR); far infrared absolute spectrophotometer (FIRAS)

Historic
Launchers

Ariane 1–4 EUROPE

Ariane 44L pictured

SPECIFICATION (Ariane 44LP):

MANUFACTURER: Aerospatiale (EADS Launch Vehicles)
LAUNCH SITE: Kourou, French Guiana
FIRST LAUNCHES: December 24, 1979 (Ariane 1); August 4, 1984 (Ariane 2–3); June 15, 1988 (Ariane 4)

FUEL:
STRAP-ONS: Solid/N_2O_4/UH25 (2)
FIRST STAGE: N_2O_4/UH25 (UDMH + 25% hydrazine hydrate)
SECOND STAGE: N_2O_4/UH25 (UDMH + 25% hydrazine hydrate)
THIRD STAGE: Liquid hydrogen/ liquid oxygen

PERFORMANCE:
PAYLOAD TO GTO: 4310 kg (10,890 lb)

PROPULSION:
SOLID STRAP-ONS: P9.5 (2)
LIQUID STRAP-ONS: Viking 6 (2)
FIRST STAGE: Viking 5 (4)
SECOND STAGE: Viking 4 (1)
THIRD STAGE: HM-7B (1)

FEATURES:
LENGTH: 58.4 m (191.6 ft)
CORE DIAMETER: 3.8 m (12.4 ft)
LAUNCH MASS: 421,000 kg (928,000 lb)

The Ariane launcher was developed by ESA and CNES in the 1970s to provide independent access to space. The three-stage Ariane 1 flew 11 times between 1979 and 1986, with nine successes. A more powerful version (Ariane 3) was introduced in 1984, with uprated engines, a stretched upper stage, and two solid propellant strap-ons. Ariane 2 was a version without the strap-ons.

The most successful member of the family was the Ariane 4, which launched half of the world's commercial satellites into GTO. It flew 116 times between 1988 and 2003, with 113 successes (a 97.4% success rate) delivering 182 satellites into orbit. Six variants of the Ariane 4 were available with different combinations of solid and liquid propellant strap-ons. The most powerful of these, the Ariane 44L, which used four liquid propellant strap-ons, could lift 4950 kg (10,890 lb) to GTO.

Payloads included one or two major satellites and six 50 kg payloads. An apogee kick motor on the satellites was used to reach GEO.

Atlas I–III USA

Atlas IIa pictured

SPECIFICATION
(Atlas IIAS):

MANUFACTURER: General
Dynamics (now Lockheed Martin)
LAUNCH SITES: Cape Canaveral,
Florida; Vandenberg, California
FIRST LAUNCH: December 16, 1993

FUEL:
STRAP-ONS: Solid
FIRST STAGE: Liquid oxygen/
kerosene (RP-1)
SECOND STAGE: Liquid hydrogen/
liquid oxygen

PERFORMANCE:
PAYLOAD TO LEO: 8610 kg
(18,980 lb)
PAYLOAD TO GTO: 3719 kg
(8200 lb)

PROPULSION:
STRAP-ONS: Thiokol Castor 4A (4)
FIRST STAGE: Rocketdyne
MA-5A (1)
SECOND STAGE: RL10A-4 or
RL10A-4-1 (2)

FEATURES:
LENGTH: 24.9 m (81.7 ft)
CORE DIAMETER: 3.3 m (10 ft)
LAUNCH MASS: 234,000 kg
(515,000 lb)

The Atlas was introduced in 1957 as America's first operational ICBM.
Despite some spectacular failures, it successfully launched four manned
Mercury spacecraft in the early 1960s. It also carried the Agena upper stage
that was used during orbital rendezvous and docking tests in the manned
Gemini program.

After the *Challenger* shuttle disaster, the Atlas II was developed in the late
1980s as an unmanned launcher for government and commercial satellites.
An improved Centaur upper stage increased its payload capability. The most
powerful version was the Atlas IIAS, with four solid rocket boosters and a
Centaur 2A upper stage. The last Atlas IIAS flew on August 31, 2004, carrying
a National Reconnaissance Office payload.

The Atlas III was introduced on May 24, 2000, during the transition to
the Atlas V. It flew six times and included the Russian RD-180 engine. This
was the first use of a Russian rocket engine to power an American launch
vehicle. The last Atlas III flew on February 3, 2005.

Black Arrow UK

SPECIFICATION:

MANUFACTURER: Saunders-Roe/
Royal Aircraft Establishment
LAUNCH SITE: Woomera, Australia
FIRST LAUNCH: June 27, 1969

FUEL:
Hydrogen peroxide, kerosene

PERFORMANCE:
PAYLOAD TO LEO: 73 kg (160 lb)

PROPULSION:
FIRST STAGE: Gamma 8 (1)
SECOND STAGE: Gamma 2 (1)
THIRD STAGE: Waxwing (1)

FEATURES:
LENGTH: 13 m (42 ft)
DIAMETER: 2 m (6.5 ft)
LAUNCH MASS: 18,130 kg
(39,960 lb)

Black Arrow was the UK's only indigenous launch vehicle. In September 1964, the government decided to develop a small, three-stage vehicle based on the Black Knight research rocket.

The first two stages used hydrogen peroxide and kerosene, topped with a Waxwing solid rocket motor manufactured by RPE Westcott. After burnout the second stage remained attached to the third stage until they coasted to apogee. Only then did the third stage spin up and separate. After a time delay, the apogee motor fired to accelerate the payload to orbital velocity.

Four Black Arrows were launched from Woomera, Australia. However, after only three test flights, one of which was successful, the project was canceled in July 1971. The fourth and final flight placed the 72.5 kg (160 lb) Prospero satellite into low Earth orbit (LEO) on 28 October 1971.

Diamant FRANCE

SPECIFICATION
(Diamant B):

MANUFACTURER: SEREB/CNES
LAUNCH SITES: Hammaguir,
Algeria; Kourou, French Guiana
FIRST LAUNCHES: November 26,
1965 (Diamant A); March 10,
1970 (Diamant B); February 6,
1975 (Diamant B P4)

FUEL:
FIRST STAGE: UDMH/N_2O_4
SECOND AND THIRD STAGES:
Solid

PERFORMANCE:
PAYLOAD TO LEO: 160 kg (350 lb)

PROPULSION:
FIRST STAGE: Valois L17 (1)
SECOND STAGE: Topaze P-2.3 (1)
THIRD STAGE: Rubis P0.6 (1)

FEATURES:
LENGTH: 23.5 m (77 ft)
DIAMETER: 1.4 m (4.6 ft)
LAUNCH MASS: 24,600 kg
(54,120 lb)

France's first satellite launcher, Diamant A, was developed as a result of the Pierres Précieuses (Precious Stones) experimental rocket program led by SEREB from 1959.

The launcher enabled France to become the third nation to place a satellite in orbit in November 1965. Diamant A's first payload was Astérix (or A-1), a small technology capsule weighing less than 40 kg. Although it was successfully placed into LEO, one of the antennas was destroyed, preventing it from communicating. Emeraude, the first stage of the Diamant A, used a liquid propulsion engine that burned for 93 sec. The second and third stages used solid propellants. The Diamant B had larger fuel tanks and a third stage derived from the Europa launcher. An uprated version was known as the B P4.

The Diamant program ended on September 27, 1975, after ten successful launches out of 12, delivering 11 satellites to orbit. All of the Diamant A launches took place from Hammaguir military base in the Algerian desert. The Diamant B and B P4 were launched from Kourou, French Guiana.

Energia (SL-17) USSR

SPECIFICATION:

MANUFACTURER: NPO Energia
LAUNCH SITE: Baikonur, Russia
FIRST LAUNCH: May 15, 1987

FUEL:
FIRST STAGE: Liquid oxygen/
kerosene
SECOND STAGE: Liquid oxygen/
liquid hydrogen

PERFORMANCE:
PAYLOAD TO LEO: 88,000 kg
(194,000 lb)
PAYLOAD TO GEO: 18,000 kg
(40,000 lb)

PROPULSION:
FIRST STAGE (STRAP-ONS):
RD-170 (4)
SECOND STAGE: RD-0120 (4)

FEATURES:
LENGTH: 58.8 m (192.9 ft)
CORE DIAMETER: 7.75 m (25.42 ft)
LAUNCH MASS: 2,524,600 kg
(5,565,700 lb)

The Energia vehicle was designed to launch large unmanned payloads and the Buran shuttle. It was the most powerful launcher in the world at the time. After 12 years of development, Energia made its maiden flight in May 1987, carrying the top-secret military Skif-DM Polyus antisatellite weapons platform. The launch was successful, but the payload failed to enter orbit due to a guidance-system failure. The second and last flight took place on November 15, 1988, when the Buran was delivered into LEO.

The strap-ons were almost identical to the first stage of the Zenit vehicle developed simultaneously by the Yuzhnoye design bureau. Designs included up to eight of these, with plans to make them reusable. Buran and other payloads were side mounted. A cryogenic upper stage was also under development. The extremely high cost of the Energia–Buran program and a shortage of potential payloads led to its cancellation in 1992. All remaining hardware was mothballed. A medium-lift Energia-M version was proposed but never flew.

Europa EUROPE

SPECIFICATION
(Europa I):

MANUFACTURER: ELDO
LAUNCH SITES: Woomera,
Australia (Europa I); Kourou,
French Guiana (Europa II)
FIRST LAUNCH: June 4, 1964

FUEL:
FIRST STAGE: Liquid oxygen/
kerosene
SECOND STAGE: UDMH/N_2O_4
THIRD STAGE: Aerozine 50/N_2O_4

PERFORMANCE:
PAYLOAD TO LEO: 1150 kg
(2530 lb)

PROPULSION:
FIRST STAGE: Rolls-Royce RZ-2 (2)
SECOND STAGE: LRBA 7T (4)
THIRD STAGE: ASAT Erno (1)

FEATURES:
LENGTH: 31.7 m (104 ft)
DIAMETER: 3.05 m (10 ft)
LAUNCH MASS: 104,670 kg
(230,750 lb)

Europa was the first launcher
developed by a collaboration of
European countries. In 1962,
Belgium, France, West Germany,
the Netherlands, Italy and the UK (associated with Australia) agreed to create
the European Launcher Development Organization (ELDO), whose main
objective was to gain independent access to space.

The ELDO A launcher (later renamed Europa I) was to launch 500–1000 kg
(1100–2200 lb) satellites into LEO. It used the British Blue Streak ballistic
missile as the first stage. France provided the Coralie second stage, and
Germany the Astris third stage. Italy was responsible for the nose fairing, and
the Netherlands and Belgium for tracking systems.

The first five tests using the Blue Streak were largely successful, but three
orbital launch attempts failed, largely due to the unreliability of the third
stage. Despite the withdrawal of the UK and Italy from the program, a
Europa II version with a solid propellant fourth stage was developed to send
small payloads into geostationary orbit. Its only flight from Kourou, on
November 5, 1971, was a failure. In 1972 the project was canceled and
replaced by Ariane.

J-1 JAPAN

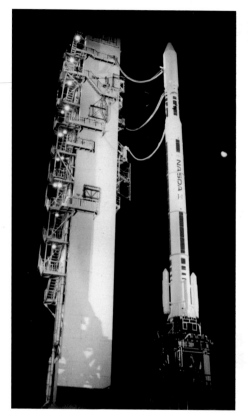

SPECIFICATION:

MANUFACTURER: Nissan
LAUNCH SITE: Tanegashima, Japan
FIRST LAUNCH: February 11, 1996

FUEL:
ALL STAGES: Solid (polybutadiene)

PERFORMANCE:
PAYLOAD TO LEO: 850 kg (1870 lb)

PROPULSION:
FIRST STAGE: H-II SRB-A
SECOND STAGE: M-23
THIRD STAGE: M-3B

FEATURES:
LENGTH: 33.1 m (108.6 ft)
CORE DIAMETER: 1.8 m (5.9 ft)
LAUNCH MASS: 88,500 kg
(194,700 lb)

The J-1 was a three-stage solid-fuel launch vehicle designed to meet growing demand for the launch of smaller satellites. It was based on the Mu-3S-II rocket developed by the Institute of Space and Astronautical Science (ISAS), a division of JAXA. The J-1 flew only once, in 1996, in a partial configuration, to launch the Hyflex (Hypersonic Flight Experiment) demonstrator.

A lower-cost version, known as the J-1 F2, was developed. It combined the solid rocket booster of the H-II with the second and third stages of the J-1, and used updated avionics. It was intended to launch the OICETS (Optical Inter-orbit Communications Engineering Test Satellite) satellite, but J-1 development ceased due to budget overruns before it could fly and the satellite was eventually launched on a Russian Dnepr.

M-V JAPAN

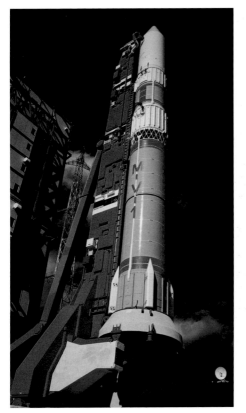

SPECIFICATION:

MANUFACTURER: Nissan
LAUNCH SITE: Uchinoura Space
Center, Kagoshima, Japan
FIRST LAUNCH: December 2, 1997

FUEL:
ALL STAGES: Solid (3 stages)

PERFORMANCE:
PAYLOAD TO LEO: 1800 kg
(3900 lb)

PROPULSION:
FIRST STAGE: M-14 (1)
SECOND STAGE: M-24 (1)
THIRD STAGE: M-34 (1)

FEATURES:
LENGTH: 30.7 m (101 ft)
DIAMETER: 2.5 m (8.2 ft)
LAUNCH MASS: 137,500 kg
(303,100 lb)

The M-V was the last in the Mu series of small satellite launchers that was flown from 1970 by Japan's Institute of Space and Astronautical Science (ISAS). One of the largest solid-fuel rockets in the world, the M-V was a three-stage vehicle with an optional kick stage. It was sufficiently powerful to launch scientific satellites such as Muses-B into Earth orbit and Nozomi to Mars. The final M-V rocket launched the Hinode satellite into Sun-synchronous orbit on September 22, 2006. Its launch record was six successes and one failure.

In an effort to fast-track the Japanese space program, the N-I and N-II (the "N" prefix is taken from the first letter of "Nippon" [Japan]) rockets were based on US technology. They were essentially versions of the Delta rocket built under license in Japan using American and Japanese components and used to launch NASDA (National Space Development Agency) satellites. The first-stage engine on the N-I was licensed from Rocketdyne, who also assisted in the development of the rocket's second-stage engine. The strap-ons, solid propellant third stage, and control systems were purchased from the US.

The N-I launched seven satellites in 1975–82 and could place up to 130 kg (286 lb) in geo-stationary orbit. The larger N-II used nine strap-ons instead of three. It had more powerful motors and an inertial guidance system. It could lift 350 kg (770 lb) to geostationary orbit and launched eight satellites in 1980–86.

SPECIFICATION (N-II):

MANUFACTURER: Mitsubishi Heavy Industries
LAUNCH SITE: Tanegashima, Japan
FIRST LAUNCH: September 9, 1975 (N-I); November 2, 1981 (N-II)

FUEL:
STRAP-ONS: Solid
FIRST STAGE: Liquid oxygen/kerosene
SECOND STAGE: N_2O_4/UDMH
THIRD STAGE: Solid

PERFORMANCE:
PAYLOAD TO GTO: 715 kg (1575 lb)

PROPULSION:
STRAP-ONS: Castor II (9)
FIRST STAGE: MB-3 (1)
SECOND STAGE: AJ-10 (1)
THIRD STAGE: TE-M-364

FEATURES:
LENGTH: 35.4 m (116 ft)
DIAMETER: 2.4 m (8 ft)
LAUNCH MASS: 135,200 kg (297,440 lb)

N-1 (G-1e/SL-15) USSR

SPECIFICATION:

MANUFACTURER: OKB-1
(Korolev/NPO Energia)
LAUNCH SITE: Baikonur, Russia
FIRST LAUNCH: February 21, 1969

FUEL:
ALL STAGES: Liquid oxygen/
kerosene (5 stages)

PERFORMANCE:
PAYLOAD TO LEO: 70,000 kg
(154,000 lb)

PROPULSION:
FIRST STAGE: NK-15 (30)
SECOND STAGE: NK-15V (8)
THIRD STAGE: NK-21 (4)
FOURTH STAGE: NK-19 (1)
FIFTH STAGE: RD-58 (1)

FEATURES:
LENGTH: 105 m (344 ft)
BASE DIAMETER: 17 m (55 ft)
LAUNCH MASS: 2,735,000 kg
(6,029,000 lb)

The N-1 was intended to send manned spacecraft to the Moon and deliver very heavy payloads to LEO. The program was given the official go-ahead in 1962, with the start of the L3 lunar-landing project in 1963.

There were four test flights of the N-1, none of which were successful. The first ended after a fire. The second began to fail immediately after liftoff when the oxidizer pump of engine 8 ingested a piece of debris and exploded. After a number of major modifications, the third launch took place on June 27, 1971. Although all 30 first-stage engines operated nominally, the rocket developed a serious roll and began to break up. It was destroyed by ground controllers after 57 sec. The final launch took place on November 23, 1972. This version included a gyrostabilized platform, steering vernier engines, and a fire-extinguishing system. The rocket flew for 107 sec without incident, but vibrations ruptured fuel lines and caused the engine 4 oxidizer pump to explode.

The next launch was scheduled for late 1974, but when Glushko became the head of NPO Energia in May 1974, he ordered the program to be terminated and all rocket stages and associated hardware to be destroyed.

Redstone (Jupiter-C/Juno) USA

SPECIFICATION
(Mercury-Redstone):

MANUFACTURER: US Army
LAUNCH SITE: Cape Canaveral,
Florida
FIRST LAUNCH: December 19,
1960

FUEL:
Liquid oxygen/ethyl alcohol

PERFORMANCE:
PAYLOAD TO 185 KM: 1300 kg
(2860 lb)

PROPULSION:
FIRST STAGE: A-6 (1)

FEATURES:
LENGTH: With Mercury, 25.3 m
(83.4 ft)
DIAMETER: 1.77 m (5.9 ft)
LAUNCH MASS: 29,395 kg
(64,669 lb)

The first Redstone was basically an enlarged version of the V-2 missile. The liquid-propelled surface-to-surface missile was developed by the Army Ballistic Missile Agency at Redstone Arsenal, Huntsville, Alabama, under the direction of Wernher von Braun. Production began in 1952. The improved Redstone 2, also named Jupiter, attained a flight range of 2400 km (1500 mi) and reached an altitude of 640 km (400 mi). A version known as the Jupiter-C was used for reentry tests of nuclear warheads. A four-stage Jupiter-C, called Juno, had upper stages comprising clusters of scaled-down Sergeant missiles. It launched America's first satellite, Explorer 1 (January 31, 1958), as well as Explorers 3 and 4. Juno II was used to launch Explorers 7, 8, and 11, and sent Pioneer 4 past the Moon into solar orbit.

About 800 modifications were made in order to use an uprated Redstone as a launcher for the first US manned missions. Three Mercury-Redstone test flights were made December 1960–March 1961. They were followed by the suborbital flights of Alan Shepard (May 5, 1961) and Gus Grissom (July 21, 1961).

Saturn IB USA

SPECIFICATION:

MANUFACTURER: NASA/Chrysler
(S-IB)/Douglas (S-IVB)
LAUNCH SITES: Cape Canaveral,
Florida; Kennedy Space Center,
Florida
FIRST LAUNCH: February 26, 1966

FUEL:
FIRST STAGE: Liquid oxygen/RP-1
SECOND STAGE: Liquid oxygen/
liquid hydrogen

PERFORMANCE:
PAYLOAD TO LEO: 15,300 kg
(33,700 lb)

PROPULSION:
FIRST STAGE: H-1 (8)
SECOND STAGE: Rocketdyne
J-2 (1)

FEATURES:
LENGTH: 68 m (224 ft)
DIAMETER: 6.6 m (21.6 ft)
LAUNCH MASS: 589,770 kg
(1,300,220 lb)

The Saturn series of rockets was proposed as a follow-on to the Jupiter series. Elements of the Saturn I and planned Saturn V were combined to build the mid-range Saturn IB. A much more powerful second stage gave the Saturn IB a 50% greater lift capability than Saturn I, making it possible to carry a complete Apollo spacecraft to LEO.

The first (S-IB) stage used eight uprated engines and burnt out at an altitude of 67.6 km (42 mi). The second (S-IVB) stage was identical to the third stage of the Saturn V except for the interstage adapter. It had a single engine that burned 242,250 l (63,996 gal) of liquid hydrogen and 75,700 l (19,998 gal) of liquid oxygen in 450 sec of operation before reaching orbital altitude.

The Saturn IB flew three test flights in 1966, two of which were suborbital demonstrations of the Apollo Command and Service Modules (CSM). The Apollo 5 flight on January 22, 1968, was an unmanned orbital test of the lunar module. On October 11, 1968, the Saturn IB launched the Apollo 7 crew into LEO. It carried three crews to the Skylab space station in 1973. The Saturn IB's final launch on July 15, 1975, delivered the Apollo 18 CSM to LEO for the Apollo–Soyuz Test Project. All of its nine flights were successful. Of the 12 launchers built, three were unused.

Saturn V USA

SPECIFICATION:

MANUFACTURER: NASA/Boeing (S-IC)/North American (S-II)/ Douglas (S-IVB)
LAUNCH SITE: Kennedy Space Center, Florida
FIRST LAUNCH: November 9, 1967

FUEL:
FIRST STAGE: Liquid oxygen/ kerosene
SECOND STAGE: Liquid oxygen/ liquid hydrogen
THIRD STAGE: Liquid oxygen/ liquid hydrogen

PERFORMANCE:
PAYLOAD TO LEO: 118,000 kg (260,000 lb)
TRANSLUNAR PAYLOAD: 47,000 kg (103,000 lb)

PROPULSION:
FIRST STAGE: F-1 (5)
SECOND STAGE: J-2 (5)
THIRD STAGE: J-2 (1)

FEATURES:
LENGTH: With Apollo, 111 m (364.1 ft); With Skylab, 109 m (357.5 ft)
DIAMETER: 10.1 m (33.1 ft)
LAUNCH MASS: Apollo 11, 2,938,312 kg (6,464,286 lb)

Saturn V, the most powerful launch vehicle ever built, was developed solely to launch Apollo crews to the Moon. It comprised completely new first and second stages, together with the S-IVB third stage used by the Saturn IB.

The five engines of the first stage produced 3.4 million kg (7.5 million lb) of thrust. They burned for 2.5 min, consuming 2 million kg (4.5 million lb) of propellant. The second stage engines burned for about 6 min, after which the vehicle had reached near-orbital velocity of 24,622 km/h (15,300 mph). The third stage engine burned for 2.75 min, boosting the spacecraft to orbital velocity. The Apollo spacecraft entered LEO about 12 minutes after liftoff. After the Apollo CSM docked with the LM, the third stage was abandoned and allowed to impact the Moon.

The Saturn V production line was shut down in 1970. Between 1967 and 1973, 13 rockets were launched with a perfect launch record, although there were engine failures on Apollo 6 and Apollo 13. Nine Saturn Vs delivered manned Apollo spacecraft to the Moon. The final launch on May 14, 1973, placed the Skylab space station in LEO.

Scout USA

SPECIFICATION
(Scout G):

MANUFACTURER: LTV Aerospace Corporation
LAUNCH SITES: Wallops Island, Virginia; Vandenberg, California; San Marco platform, Kenya
FIRST LAUNCHES: July 1, 1960 (Scout X-1); October 30, 1979 (Scout G)

FUEL:
ALL STAGES: Solid

PERFORMANCE:
PAYLOAD TO LEO: 210 kg (460 lb)

PROPULSION:
FIRST STAGE: Algol 3 (1)
SECOND STAGE: Castor 2 (1)
THIRD STAGE: Antares 3A (1)
FOURTH STAGE: Altair 3 (1)

FEATURES:
LENGTH: 26 m (85 ft)
DIAMETER: 1.1 m (3.7 ft)
LAUNCH MASS: 20,930 kg (46,140 lb)

Scout was developed in the late 1950s as the first US solid-propellant rocket capable of placing payloads in orbit. It was a reliable, versatile, and cost-effective vehicle that was used in nine configurations over 34 years. It could launch suborbital probes and high-speed reentry vehicles as well as small orbital payloads.

Early Scouts were capable of boosting less than 150 kg (330 lb) into a nominal 500 km orbit, but performance was enhanced by changing to larger, more powerful solid motors. The original first stage evolved from the Polaris missile, the second stage was derived from the Sergeant missile, and the fourth stage was derived from the Vanguard program. A fifth stage could be added for highly elliptical orbits or improved hypersonic reentry performance.

In addition to use by NASA and the US DOD, Scout launched payloads from the UK, Italy, France, Germany, and the European Space Research Organization (ESRO). The last launch, a ballistic missile defense satellite called MSTI, took place on May 9, 1994. There were 118 Scout launches, with an 88% success rate. Most failures occurred in the 1960s.

Sputnik/Vostok/Voskhod (A/A1) USSR

The A series of launchers was based on the R-7, the Soviet Union's first ICBM. Its first successful launch placed Sputnik, the world's first artificial satellite, into LEO. During the following years, there were some spectacular successes, but also numerous failures that were hidden from the public. The largest payload, Sputnik 3, had a mass of 1327 kg (2925 lb).

The Sputnik launcher had a central core with four cone-shaped liquid strap-on boosters. The core engine ignited simultaneously with the strap-ons, so at liftoff the vehicle had 20 main chambers and 12 vernier chambers in operation.

The more powerful Vostok variant was introduced in 1959 to send the first three Luna spacecraft to the Moon. This version had a second stage (stage E) and a larger payload fairing that was used for six manned Vostok missions and their precursor test flights. For the manned Voskhod missions, the upper-stage propellant tanks were stretched and a more powerful engine added. The first commercial launch of a Vostok, with the Indian IRS-1A satellite, took place on March 17, 1988. Its last launch was on August 29, 1991.

SPECIFICATION
(Vostok):

MANUFACTURER: Korolev Design Bureau (OKB-1)
LAUNCH SITES: Baikonur, Kazakhstan; Plesetsk, Russia
FIRST LAUNCH: October 4, 1957 (Sputnik); May 15, 1960 (Vostok)

FUEL:
ALL STAGES: Liquid oxygen/ kerosene

PERFORMANCE:
PAYLOAD TO LEO: 4730 kg (10,400 lb)

PROPULSION:
STRAP-ONS: RD-107 (4)
CORE STAGE: RD-108 (1)
SECOND STAGE: RD-448 (1)

FEATURES:
LENGTH: 38.4 m (126 ft)
CORE DIAMETER: 2.7 m (8.9 ft)
LAUNCH MASS: 290,000 kg (639,000 lb)

Thor USA

SPECIFICATION
(Thor Able):

MANUFACTURER: Douglas Aircraft Company
LAUNCH SITES: Cape Canaveral, Florida; Vandenberg, California
FIRST ORBITAL LAUNCH: August 17, 1958

FUEL:
FIRST STAGE: Liquid oxygen/RJ-l
SECOND STAGE: Nitric acid/UDMH
THIRD STAGE: Solid

PERFORMANCE:
PAYLOAD TO LEO: 120 kg (260 lb)

PROPULSION:
FIRST STAGE: LR-79 (1)
SECOND STAGE: AJ10-40 (1)
THIRD STAGE: Altair ABL 248 (1)

FEATURES:
LENGTH: 30 m (98 ft)
CORE DIAMETER: 2.4 m (8 ft)
LAUNCH MASS: 51,608 kg (113,776 lb)

Thor was the US Air Force's answer to the Jupiter missile being developed in the 1950s by the US Army and Navy. It became the first operational intermediate-range ballistic missile deployed by the US. Thor's first launches were reentry vehicle test flights, followed by the failed launch of the first Pioneer—an attempt to send a space probe to the Moon—in August 1958.

The first successful launch of the two-stage Thor-Agena-A was on February 28, 1959. It was used by the USAF to launch the first reconnaissance satellites of the Corona (Discoverer) series. The Thor-Agena-B was almost exclusively used to launch military satellites. The Thor-Agena-D, first launched on June 27, 1962, remained in service until 1968. In 1965, it carried out the first launch of a NASA satellite from Vandenberg Air Force Base.

In 1963, three solid motors derived from the Sergeant missile and an improved main engine were added to increase liftoff thrust. This version, known as the Thrust-Augmented Thor, used upper stages from the Agena-D and, later, the Delta. Stretched versions of the Thor also flew in the late 1960s. The Agena vehicle was used in large numbers during the 1960s and 1970s as the upper stage for the Atlas and Titan missiles.

Titan I–IV USA

Titan IV pictured

SPECIFICATION (Titan IV):

MANUFACTURER: Martin Marietta Astronautics
LAUNCH SITES: Cape Canaveral, Florida; Vandenberg, California
FIRST LAUNCHES: April 8, 1964 (Titan II); September 1, 1964 (Titan III); June 14, 1989 (Titan IV)

FUEL:
STRAP-ONS: Solid (2)
FIRST STAGE: Aerozine 50/N_2O_4
SECOND STAGE: Aerozine 50/N_2O_4
THIRD STAGE: Solid/HTPB (IUS) or liquid oxygen/liquid hydrogen (Centaur)

PERFORMANCE:
PAYLOAD TO LEO: with SRMU, 21,640 kg (47,700 lb)
PAYLOAD TO GTO: with SRMU, 8620 kg (19,000 lb)

PROPULSION:
STRAP-ONS: SRM (2) or SRMU (2)
FIRST STAGE: LR87 – AJ-11 (2)
SECOND STAGE: LR91 – AJ-11 (1)
IUS THIRD STAGE: SRM-1 (1) and SRM-2 (1)
CENTAUR THIRD STAGE: RL-10A (2)

FEATURES:
LENGTH: 62.2 m (204 ft)
CORE DIAMETER: 3.1 m (10 ft)
LAUNCH MASS: 860,000 kg (1.9 million lb)

The two-stage Titan II space launcher was based on the ICBM of the same name and flew 12 times in the mid-1960s. It was chosen for the Gemini manned launches because of its non-explosive storable propellants.

The Titan II was the basis for the many versions of its more powerful successor. The Titan IIIA, a Titan II with a new upper stage, only flew three times. The Titan IIIB was a Titan II with an Agena third stage. The IIIC comprised a IIIA plus two strap-on solid rocket motors. The IIID was a IIIC without the upper stage, while the IIIE was a IIID with a powerful liquid-fuel Centaur upper stage. This version was used to launch the Helios, Viking, and Voyager interplanetary spacecraft.

After the loss of the *Challenger* shuttle in 1986, Martin Marietta used its own funds to develop the commercial Titan III. The Titan IV, the most power-ful unmanned US rocket, was developed for the Air Force. With a Centaur upper stage and two upgraded solid rocket motors, it was used to send the Cassini spacecraft to Saturn. Its last flight was on October 19, 2005.

Vanguard USA

SPECIFICATION:

MANUFACTURER: Naval Research
Laboratory
LAUNCH SITE: Cape Canaveral,
Florida
FIRST ORBITAL LAUNCH:
December 6, 1957

FUEL:
FIRST STAGE: Liquid oxygen/
kerosene
SECOND STAGE: Nitric acid/
UDMH

PERFORMANCE:
PAYLOAD TO LEO: 25.4 kg (56 lb)

PROPULSION:
FIRST STAGE: X-405 (1)
SECOND STAGE: AJ10-118 (1)

FEATURES:
LENGTH: 23 m (75 ft)
DIAMETER: 1.1 m (3.7 ft)
LAUNCH MASS: 10,050 kg
(22,150 lb)

In 1955, the Navy's proposed Vanguard vehicle was selected by the US DOD to launch America's first satellite as part of the International Geophysical Year. Its non-missile origin was a factor in this decision. The Vanguard first stage was based on the Viking, the largest of the US liquid-fuelled scientific sounding rockets. The second stage was adapted from the Navy's Aerobee sounding rocket. The third stage used a new solid-fuel engine.

The prototype of the Vanguard flew a successful suborbital mission on October 23, 1957. The first attempt to launch a satellite took place in December 1957 but ended with the vehicle exploding on the pad. After another failure in February 1958, the vehicle successfully launched the 2 kg (4.4 lb) Vanguard 1—America's second satellite—on March 17, 1958. Two more small satellites were launched before Vanguard was retired. Altogether, there were eight failures in 12 Vanguard launches. The final launch took place on September 18, 1959. Despite its only modest success, the Vanguard upper stages led to the Able upper stage for Thor and Atlas.

Current/
Future
Launchers

Angara RUSSIA

Angara 1.1 pictured

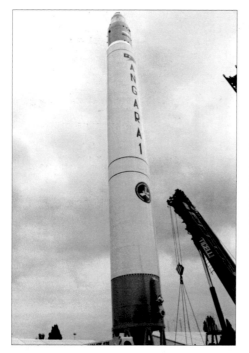

SPECIFICATION:
(Angara 1.1)

MANUFACTURER: Khrunichev
LAUNCH SITE: Plesetsk, Russia
FIRST LAUNCH: 2011?

FUEL:
FIRST STAGE: Liquid oxygen/
kerosene
SECOND STAGE: Breeze M,
N_2O_4/UDMH

PERFORMANCE:
PAYLOAD TO LEO: 2000 kg
(4400 lb)

PROPULSION:
FIRST STAGE: RD-191 (1)
SECOND STAGE: Breeze M,
RD-2000 (1)

FEATURES:
LENGTH: 34.9 m (114.5 ft)
CORE DIAMETER: 3.9 m (12.7 ft)
LAUNCH MASS: 145,000 kg
(319,000 lb)

The Angara program began in 1993 when the Russian Ministry of Defense and the Russian Federal Space Agency announced a tender to develop a new national space launch system. Khrunichev was eventually named as the prime contractor. Although the program has received financial support from the federal budget, the rocket's maiden flight has been delayed several years.

Khrunichev planned to produce light, medium, and heavy versions, based on a Multi-Purpose Rocket Module (MPRM) that uses the RD-191 engine. The Angara 1.1 and Angara 1.2 will use one MPRM as a first stage, while the central part of the Breeze-M upper stage and the Block 'I' of the Soyuz-2 will form the second stage. This will be able to launch payloads of 2–3.7 t (4400–8140 lb) to LEO. The medium-class Angara 3A will have two additional, side-mounted MPRMs, while the heavy Angara 5A will have five MPRMs as its first stage. The eventual goal is to replace the Proton launcher with Angara 5A at Baikonur and Plesetsk. All members of the Angara family will be launched from the same pads.

Ares I (Crew Launch Vehicle/CLV) USA

SPECIFICATION:

MANUFACTURER: ATK Thiokol, Boeing Space Exploration
LAUNCH SITE: Kennedy Space Center, Florida
FIRST LAUNCH: 2009?

FUEL:
FIRST STAGE: Solid
SECOND STAGE: Liquid oxygen/ hydrogen

PERFORMANCE:
PAYLOAD TO LEO: 25,000 kg (55,000 lb)

PROPULSION:
FIRST STAGE: Shuttle SRB (5)
SECOND STAGE: J-2X (1)

FEATURES:
LENGTH: 93 m (309 ft)
DIAMETER: 10 m (33 ft)
LAUNCH MASS: 900,000 kg (1.8 million lb)

The Ares I will launch manned and cargo versions of the Orion spacecraft after the end of the Space Shuttle flight program in 2010. In the current design, Ares I will use a five-segment version of the Space Shuttle Solid Rocket Booster (SRB) for its first stage, with a liquid-fuel second stage powered by a single J-2X rocket engine derived from the Saturn rockets. The Ares I will also be used to launch the unmanned cargo supply version of the Orion spacecraft. Orion, like its Apollo predecessor, will be mounted on the top of the rocket, eliminating the possibility of damage by debris shaken from the launch vehicle.

The first suborbital test flight (called Ares I-X) is scheduled for 2009. This will test a first stage comprising four active SRB segments with an inert fifth segment and upper stage. A test flight of the fully operational Ares I launcher is scheduled for 2013, followed by an orbital test with the fully operational Orion spacecraft. The initial crewed launch of the complete Ares I/Orion vehicle should follow around 2015. The larger, unmanned Ares V cargo launch vehicle will be used to launch the Lunar Surface Access Module into LEO for later retrieval by an Orion crew.

Ariane 5 EUROPE

Ariane 5 ECA pictured

SPECIFICATION (Ariane 5 ECA):

MANUFACTURER: EADS Space Transportation
LAUNCH SITE: Kourou, French Guiana
FIRST LAUNCH: June 4, 1996 (Ariane 5 dev.); December 11, 2002 (Ariane 5 ECA)

FUEL:
STRAP-ONS: Solid
FIRST STAGE: Liquid oxygen/liquid hydrogen
SECOND STAGE: Liquid oxygen/liquid hydrogen

PERFORMANCE:
PAYLOAD TO GTO: 9600 kg (21,120 lb)

PROPULSION:
STRAP-ONS: EAP
FIRST STAGE: Vulcain 2 (1)
SECOND STAGE: HM7B (1)

FEATURES:
LENGTH: 50.5–57.7 m (165.6–189.3 ft)
CORE DIAMETER: 5.4 m (17.7 ft)
LAUNCH MASS: 780,000 kg (1.7 million lb)

Since 2003, ESA has been in charge of Ariane program management, while EADS Space Transportation has been prime contractor. ESA and CNES were responsible for the first three Ariane 5 launches, then Arianespace became responsible for commercial operations.

Ariane 5 features a radically new design from its predecessors. Originally intended to carry Europe's manned Hermes minishuttle, it is now solely a commercial launcher. The basic Ariane 5G had a payload capacity of 6900 kg (15,180 lb) for dual launches into GTO. This has been phased out in favor of the Ariane 5 ECA (Evolution Cryogénique A), which incorporates a new cryogenic upper stage carrying over 14 t (30,865 lb) of propellants. Up to eight secondary payloads can be carried with an ASAP (Ariane Structure for Auxiliary Payloads) platform.

Future versions include the Ariane 5 ES (Evolution Stockables) and the Ariane 5 ECB (Evolution Cryogénique B). The ES will deliver the Automated Transfer Vehicle to the International Space Station from 2007. The ECB will use the new Vinci multiple-ignition engine and be able to launch up to 12 t (26,400 lb) into GTO.

Atlas V <small>USA</small>

SPECIFICATION
(Atlas V 551):

MANUFACTURER: Lockheed Martin
LAUNCH SITES: Cape Canaveral,
Florida; Vandenberg, California
FIRST LAUNCH: August 21, 2002
(Atlas V 401)

FUEL:
STRAP-ONS: Solid
FIRST STAGE: Liquid oxygen/RP1
SECOND STAGE: Centaur—liquid
oxygen/liquid hydrogen

PERFORMANCE:
PAYLOAD TO LEO: 18,500 kg
(40,780 lb)
PAYLOAD TO GTO: 8700 kg
(19,180 lb)

PROPULSION:
STRAP-ONS: Aerojet
FIRST STAGE: RD-180 (1)
SECOND STAGE: RL10A-4-2
(1 or 2)
LENGTH: 59.7 m (196 ft)
CORE DIAMETER: 3.8 m (12.5 ft)
LAUNCH MASS: 333,298 kg
(734,800 lb)

The Atlas V was developed as part of the US Air Force Evolved Expendable Launch Vehicle (EELV) program. There are six versions of the Atlas V 500 series, of which the Atlas V 551 is the most powerful. The numbers in the designation refer to the fairing diameter (m), the number of solid rocket boosters, and the number of Centaur engines. The Atlas V 401 has the smallest fairing.

Atlas V 500 series launchers feature a Common Core Booster first stage, equipped with a Russian-developed RD-180 engine, plus zero to five strap-on solid rocket boosters. Their second stage is the multiburn Centaur with one or two high-performance engines. The Atlas V used for NASA's New Horizons mission also carried a custom-built Boeing solid-propellant Star 48B third-stage motor to send the spacecraft on its way to Pluto at a record escape velocity. New Horizons left the Earth at approximately 57,600 km/h (36,000 mph).

Delta II USA

SPECIFICATION
(Delta II 7000 series):

MANUFACTURER: Boeing
LAUNCH SITES: Cape Canaveral,
Florida; Vandenberg, California
FIRST LAUNCHES: May 13, 1960
(Delta); February 14, 1989 (Delta
II); August 27, 1998 (Delta III)

FUEL:
STRAP-ONS: Solid
FIRST STAGE: Liquid oxygen/
kerosene
SECOND STAGE: N_2O_4/aerozine
(UDMH and hydrazine)
THIRD STAGE: (PAM-D) solid

PERFORMANCE:
PAYLOAD TO LEO: 2700-6100 kg
(5960-13,440 lb)
PAYLOAD TO GTO: 900-2170 kg
(1980 1790 lb)

PROPULSION:
STRAP-ONS: GEM 40 (3-9)
FIRST STAGE: Rocketdyne
RS-27A (1)
SECOND STAGE: Aerojet AJ10-
118K (1)
THIRD STAGE: Thiokol Star 48B
solid rocket motor (1)

FEATURES:
LENGTH: 38.2–39 m (125.3–27 ft)
DIAMETER: 2.4 m (8 ft)
LAUNCH MASS: 151,700–231,870 kg
(334,300–511,180 lb)

The original Delta was derived from the US Air Force Thor intermediate-range ballistic missile. The vehicle was modified during the 1960s and 1970s so that it could use two or three stages, augmented with three, six, or nine solid-fuel strap-ons. NASA's first successful Delta launch was the Echo 1A satellite on August 12, 1960, and the Delta remained NASA's primary launcher until the advent of the Shuttle halted production in 1981–86.

Following the *Challenger* shuttle disaster, the US Air Force contracted to build a more powerful Delta II, which was introduced in 1989. The Delta II can be configured as a two-stage or three-stage launcher, with a varying number of strap-on solid rocket boosters and two sizes of payload fairings. For example, a Delta II 7925 has the 7000-series first stage, nine strap-ons, a second stage, and a Star 48B third stage. The Delta II Heavy has the larger GEM-46 solid rocket boosters. Development of the more powerful Delta III commercial launcher began in 1995, with first launch in 1998. Of three flights, two were failures, the third carried a dummy payload. It was replaced by the Delta IV.

Delta IV USA
Delta IV Heavy pictured

SPECIFICATION
(Delta IV Heavy):

MANUFACTURER: Boeing
LAUNCH SITES: Cape Canaveral,
Florida; Vandenberg, California
FIRST LAUNCH: December 21,
2004

FUEL:
FIRST STAGE: Liquid oxygen/
liquid hydrogen
SECOND STAGE: Liquid oxygen/
liquid hydrogen

PERFORMANCE:
PAYLOAD TO LEO: 23,040 kg
(50,800 lb)
PAYLOAD TO GTO: 13,130 kg
(28,950 lb)

PROPULSION:
FIRST STAGE: Rocketdyne
RS-68 (3)
SECOND STAGE: Pratt & Whitney
RL10B-2 (1)

FEATURES:
LENGTH: 70.7 m (231.9 ft)
CORE DIAMETER: 5 m (16.4 ft)
LAUNCH MASS: 733,400 kg
(1.6 million lb)

The Delta IV family of medium- to heavy-lift launch vehicles was developed in the 1990s under the US Air Force Evolved Expendable Launch Vehicle (EELV) program. There are five launch vehicle configurations based on a new liquid oxygen/hydrogen Common Booster Core (CBC) first stage. Its RS-68 motor was the first new cryogenic engine produced in the US in over 20 years.

A Delta IV Small launcher was proposed without additional strap-ons, but it has never flown. The Delta IV Medium consists of a single CBC first stage and a second stage fitted with the Pratt & Whitney RL10B-2 engine. There is a choice of two sizes of expanded fuel and oxidizer tanks. The Delta IV Medium+ has three variants, each using a single CBC augmented by either two or four solid rocket strap-on graphite epoxy motors (GEMs). It can carry a 4 m (13.1 ft) or 5 m (16.4 ft) diameter fairing.

The most powerful version, the Delta IV Heavy, uses two CBCs on either side of another CBC, plus a single cryogenic second stage. During its first flight in December 2004, a demonstration satellite was deployed in a lower-than-expected orbit.

Dnepr (SS-18/Satan, R-36M) RUSSIA

SPECIFICATION:

MANUFACTURER: Yuzhnoe Design Bureau
LAUNCH SITES: Baikonur, Russia; Dombarovsky/Yasnyy, Russia
FIRST LAUNCH: April 21, 1999

FUEL:
ALL STAGES: N_2O_4/UDMH

PERFORMANCE:
PAYLOAD TO LEO: 4500 kg (9900 lb)

PROPULSION:
FIRST STAGE: RD-264 (1)
SECOND STAGE: RD-0255 (1)
THIRD STAGE: RD-869 (1)

FEATURES:
LENGTH: 34.3 m (113.2 ft)
DIAMETER: 3 m (9.8 ft)
LAUNCH MASS: 211,000 kg (465,000 lb)

The Dnepr is a modified version of the Russia–Ukraine SS-18 ICBM, the world's most powerful ballistic missile. Following the START (Strategic Arms Reduction) treaty, at least 115 SS-18 missiles were slated for conversion into space launch vehicles. In 1997, International Space Company Kosmotras was established to manage and market the launcher program.

Like other Russian ICBMs, the rocket is fuelled during production and leaves the manufacturer's premises ready for launch. It uses toxic liquid propellants, which have been criticized for their environmental impact, but the missile can be stored for years without performance degradation.

The first commercial launch took place from Baikonur on April 21, 1999, using a missile that had been on duty for over 20 years, to launch the UK-built UoSat-12. After this success, the guaranteed service life of this type was extended to at least 25 years.

The first space launch of a Dnepr from Yasnyy (Dombarovsky ICBM base), took place on July 12, 2006, when it delivered the Genesis I inflatable spacecraft into orbit. ISC Kosmotras plans to continue Dnepr launches until 2017, when the guaranteed service life expires. During that period the Dnepr's reliability will be checked annually.

Falcon USA

SPECIFICATION:

MANUFACTURER: SpaceX
LAUNCH SITES: Vandenberg, California; Cape Canaveral, Florida; Kwajalein, Marshall Islands; Kodiak, Alaska; Wallops, Virginia
FIRST LAUNCH: March 25, 2006

FUEL:
FIRST STAGE: Liquid oxygen/RP-1 (refined petroleum)
SECOND STAGE: Liquid oxygen/RP-1 (refined petroleum)

PERFORMANCE:
PAYLOAD TO LEO: 670 kg (1480 lb)

PROPULSION:
FIRST STAGE: SpaceX Merlin (1)
SECOND STAGE: SpaceX Kestrel (1)

FEATURES:
LENGTH: 21.3 m (70 ft)
DIAMETER: 1.7 m (5.5 ft)
LAUNCH MASS: 27,200 kg (60,000 lb)

Falcon 1 is a privately funded launcher being developed by SpaceX as a low-cost means of access to LEO for commercial payloads. The first stage returns by parachute for a water landing and is recovered for reuse, while the second stage is not reusable.

After lengthy delays, the first Falcon 1 launcher lifted off from Kwajalein Atoll in the Marshall Islands on March 25, 2006. On board was the US Air Force Academy's FalconSAT–2. A fuel leak caused a fire in the first stage and the rocket crashed onto a nearby reef. The second flight was launched from Kwajalein on March 20, 2007. The rocket failed to reach orbit. The next was expected to carry the US Naval Research Laboratory's Tacsat 1 and a Space Services Incorporated payload of cremated human remains.

Falcon 5 and Falcon 9 will be larger, more powerful versions with five and nine Merlin engines respectively. They will use the same structural architecture, avionics, and launch system as the Falcon 1, but both stages will be reusable. SpaceX hopes to use Falcon 9 to fly its reusable Dragon capsule to the International Space Station.

Geostationary Launch Vehicle (GSLV) INDIA

SPECIFICATION:

MANUFACTURER: ISRO
LAUNCH SITES: Sriharikota, India
FIRST LAUNCH: April 18, 2001

FUEL:
FIRST STAGE: Strap-ons, N_2O_4/
UDMH; core, solid
SECOND STAGE: N_2O_4/UDMH
THIRD STAGE: Liquid oxygen/
liquid hydrogen

PERFORMANCE:
PAYLOAD TO LEO: 5000 kg
(11,000 lb)
PAYLOAD TO GTO: 2500 kg
(5500 lb)

PROPULSION:
FIRST STAGE: Strap-ons, Vikas L40
(4); core, PSLV-1 (1)
SECOND STAGE: Vikas L37.5 (1)
THIRD STAGE: RD-56M (1)

FEATURES:
LENGTH: 49 m (160 ft)
CORE DIAMETER: 2.8 m (9.1 ft)
LAUNCH MASS: 402,000 kg
(886,000 lb)

The GSLV was developed to launch Indian and foreign communication satellites of 2000–2500 kg (4400–5500 lb) into GTO. It uses major components from the PSLV launcher, particularly the core stage solid booster and the Vikas liquid-fuel engine. These Indian-built Vikas L-40 motors are based on the Ariane Viking-2 engine.

The first stage core motor, which is made up of five segments, carries 129–138 t (284.4–304.2 lb) of HTPB. It is one of the largest solid propellant boosters in the world. The core stage is ignited 4.6 sec after confirming the normal operation of each of the L40 strap-ons. The second stage uses a Vikas engine based on the French Viking-4A. The upper stage uses a cryogenic engine from Russia; this will eventually be replaced by a more powerful cryogenic stage.

The development of the GSLV Mk-III was approved by the Indian government in April 2002. This is an entirely new three-stage launcher. It will initially be able to launch a 4400 kg (9680 lb) satellite to GTO, or 10,000 kg (22,000 lb) to LEO, with growth potential towards a 6000 kg (13,200 lb) payload capability.

H-IIA JAPAN

The H-I was the first Japanese launcher to incorporate an indigenously developed liquid oxygen/liquid hydrogen (LOX/LH) engine. Nine satellites were launched by the H-I in 1986-1991. It was replaced by the H-II rocket, which had two stages powered by LOX/LH. However, it was very expensive to launch and suffered two launch failures in 1999.

The H-IIA, Japan's current heavy launch vehicle, has modified engines, fuel tanks, strap-on boosters, and guidance systems. Introduced in 2001, it has a similar launch capability to the H-II, but it is cheaper and more reliable to operate. Three versions are currently in use. The basic 202 vehicle has two LOX/LH stages with two SRB-A solid rocket boosters strapped to the side. The most powerful version is the 204, which has four SRB-As.

In addition to launches to GTO, the H-IIA has delivered a payload of four satellites to polar orbit. A yet-to-be flown "Augmented" H-IIB version with a wider first stage and two LE-7A engines is being developed. This will be able to deliver Japan's H-II Transfer Vehicle to the International Space Station.

SPECIFICATION:

MANUFACTURER: Mitsubishi Heavy Industries
LAUNCH SITE: Tanegashima, Japan
FIRST LAUNCH: August 29, 2001

FUEL:
STRAP-ONS: Solid
FIRST STAGE: Liquid oxygen/liquid hydrogen
SECOND STAGE: Liquid oxygen/liquid hydrogen

PERFORMANCE (HIIA-2024):
PAYLOAD TO GTO: 5000 kg (11,000 lb)

PROPULSION:
STRAP-ONS: SRB-A (2), SSB (4)
FIRST STAGE: LE-7A (1)
SECOND STAGE: LE-5B (1)

FEATURES:
LENGTH: 52.5 m (172.2 ft)
CORE DIAMETER: 4 m (13.1 ft)
LAUNCH MASS: 351,000 kg (772,000 lb)

Kosmos 3M (Cosmos, SL-8 or C-1) RUSSIA

SPECIFICATION:

MANUFACTURER: Polyot Design
Bureau
LAUNCH SITES: Plesetsk, Russia;
Kapustin Yar, Russia
FIRST LAUNCH: August 18, 1964

FUEL:
ALL STAGES: N_2O_4/UDMH

PERFORMANCE:
PAYLOAD TO LEO: 1400 kg
(3080 lb)

PROPULSION:
FIRST STAGE: RD-216 (2)
SECOND STAGE: RD-219 (1)

FEATURES:
LENGTH: 32.4 m (106.3 ft)
CORE DIAMETER: 2.4 m (7.9 ft)
LAUNCH MASS: 109,000 kg
(239,800 lb)

Kosmos is a two-stage liquid-fuelled rocket developed in the 1960s by the
Yangel Design Bureau. Kosmos 1, with a first stage based on the R-14
medium-range missile, was launched from Baikonur in 1964. An improved
version (Kosmos 3) was in use until 1977, delivering 500 kg (1100 lb) into
LEO, including payloads such as the Kosmos and Intercosmos satellites.
There were also hundreds of high velocity reentry tests.

The Kosmos 3M, a more powerful version with an upgraded second
stage, was developed in the late 1960s. In 1968, the production of this vehicle
was transferred to PO Polyot in the eastern Siberian town of Omsk. The 3M
has delivered eight small satellites into orbit, and, between 1980 and 1988,
launched 10 BOR scaled prototypes to test the aerodynamics of the future
Buran shuttle orbiter. Polyot continued to manufacture the Kosmos 3M until
1994, when production was suspended. It has now resumed. In October 2005,
the 3M successfully orbited the SSETI Express student satellite. It is also
scheduled to orbit five German SAR-Lupe satellites between 2006 and 2008.

Long March (Chang Zheng/CZ) CHINA

(LM-3B shown)

SPECIFICATION
(Long March 3B):

MANUFACTURER: CALT
LAUNCH SITES: Xichang, China
(LM-2E, 3, 3B); Juiquan, China
(LM-2C, 2D, 2F); Taiyuan, China
(LM-4)
FIRST LAUNCH: February 14, 1996

FUEL:
STRAP-ONS: N₂O₄/UDMH
FIRST STAGE: N₂O₄/UDMH
SECOND STAGE: N₂O₄/UDMH
THIRD STAGE: Liquid hydrogen/
liquid oxygen

PERFORMANCE:
PAYLOAD TO LEO: 11,200 kg
(24,600 lb)
PAYLOAD TO GTO: 5100 kg
(11,200 lb)

PROPULSION:
STRAP-ONS: YF-20 (4)
FIRST STAGE: YF-21 (1)
SECOND STAGE: YF-22 (1)
THIRD STAGE: YF-75 (1)

FEATURES:
LENGTH: 54.8 m (179.7 ft)
CORE DIAMETER: 3.4 m (11 ft)
LAUNCH MASS: 425,800 kg
(938,700 lb)

Over the last 40 years, China has developed a series of Long March launch vehicles that offer a wide range of performance to meet the demands of government and commercial payloads. The LM-2 series, generally used for LEO missions, includes the LM-2E, which can also lift 3500 kg (7700 lb) into GTO, and the LM-2F, which has been developed for the crewed Shenzhou spacecraft. The LM-3 series is used for launches of communications satellites to GTO, and the LM-4 is used for various missions to polar- or Sun-synchronous orbit. An LM-5 version is under development and expected to eventually replace the current family.

The LM-3B is the most powerful version currently available for GTO missions. It is based on the LM-3A core stage but has four liquid strap-on boosters—identical to those on the LM-2E—and enlarged second-stage fuel tanks.

The first LM-3B launch was a failure. Since then, all launches have been successful. Sanctions and technology-transfer restrictions by the USA meant that there were no LM-3B launches between 1998 and 2005.

Minotaur USA

SPECIFICATION:

MANUFACTURER: Orbital Sciences
Corporation
LAUNCH SITES: Vandenberg,
California; Cape Canaveral,
Florida; Wallops, Virginia;
Kodiak Island, Alaska
FIRST LAUNCH: January 27, 2000

FUEL:
ALL STAGES: Solid (4 stages)

PERFORMANCE:
PAYLOAD TO LEO: 580 kg (1280 lb)

PROPULSION:
FIRST STAGE: M55A1 (1)
SECOND STAGE: SR19 (1)
THIRD STAGE: Orion 50XL (1)
FOURTH STAGE: Orion 38 (1)

FEATURES:
LENGTH: 19.2 m (63 ft)
DIAMETER: 1.7 m (5.5 ft)
LAUNCH MASS: 36,200 kg
(79,800 lb)

The Minotaur I is a small launcher developed under the US Air Force's Orbital/Suborbital Program to utilize surplus Minuteman II missiles. It uses Minuteman rocket motors for the first and second stages, reusing engines decommissioned as a result of arms-reduction treaties, combined with the upper stages and payload fairing of the Pegasus XL. Its capabilities have been enhanced with improved avionics systems. On its maiden launch (from Vandenberg), it delivered 11 US nanosatellites to LEO. The first launch from Wallops took place on December 16, 2006.

The more powerful Minotaur IV is currently under development. It combines the solid rocket motors from decommissioned Peacekeeper ICBMs with technologies from other Orbital-built launch vehicles, including the Minotaur I, Pegasus, and Taurus. The first Minotaur IV is scheduled to launch the US Air Force's Space-Based Surveillance System (SBSS) satellite in December 2008. A Minotaur V five-stage version for GTO and interplanetary missions is also under study.

Pegasus XL ^{USA}

Funded and developed by Orbital Sciences Corporation, Pegasus was the first all-new US launcher since the 1970s and the world's first privately developed space launch vehicle. It was designed as a low-cost, three-stage, solid-propellant vehicle to carry small satellites to a variety of LEOs.

During its first six missions, Pegasus was air-launched from a NASA B-52 aircraft flying from Dryden Flight Research Center, California. Orbital then began to use a modified Stargazer L-1011 aircraft. An uprated version, the Pegasus XL, was introduced in 1994. Aerodynamic lift generated by its unique delta-shaped wing enables Pegasus to deliver satellites to orbit in a little over 10 minutes.

The standard three-stage Pegasus has been used to launch multiple payloads, notably eight Orbcomm satellites. There is also an optional Hydrazine Auxiliary Propulsion System fourth stage, first flown in 1997. A wingless version, called the Orbital Boost Vehicle, has been developed to support missile defense technology development.

SPECIFICATION:

MANUFACTURER: Orbital Sciences Corporation
LAUNCH SITES: (Air launched) Vandenberg, California; Wallops, Virginia; Kwajalein, Marshall Islands (Pegasus only); Cape Canaveral, Florida; Dryden, California (Pegasus only); Gando, Canary Islands
FIRST LAUNCH: April 5, 1990 (Pegasus); April 27, 1994 (Pegasus XL)

FUEL:
ALL STAGES: Solid (3 stages)

PERFORMANCE:
PAYLOAD TO LEO: 440 kg (968 lb)

PROPULSION:
FIRST STAGE: Orion 50SXL (1)
SECOND STAGE: Orion 50XL (1)
THIRD STAGE: Orion 38 (1)

FEATURES:
LENGTH: 17.6 m (57.7 ft)
DIAMETER: 1.3 m (4.2 ft)
LAUNCH MASS: 24,000 kg (52,000 lb)

Proton (UR-500, D-1/D-1e, SL-9, SL-12/SL-13) RUSSIA

Proton-K pictured

The Proton was developed as a heavy-lift launcher between 1961 and 1965. Its name originates from a series of large scientific satellites that were among the rocket's first payloads. It was the first Soviet launcher not based on an existing ballistic missile. The original version of the Proton (SL-9) had two stages, the first with a cluster of six engines and the second with four engines. It was only used for four launches in 1965–66.

A modified version, now known as the Proton K, was first flown in 1968. The three-stage model launched large payloads into LEO, including all Soviet space station modules. A four-stage Proton, fitted with an Energia Block D or DM upper stage, was first flown in 1967. This has been used to launch planetary missions and deliver communications satellites to GTO or GEO.

A modified version, the Proton M, introduced in 2001, uses a Khrunichev Breeze M fourth stage capable of multiple restarts. It mainly launches commercial payloads. Both the Proton K and Proton M can launch multiple payloads into LEO or GEO.

SPECIFICATION:

MANUFACTURER: Khrunichev State Research and Production Center
LAUNCH SITE: Baikonur, Kazakhstan
FIRST LAUNCHES: July 16, 1965 (Proton); March 10, 1967 (Proton K); April 7, 2001 (Proton M)

FUEL:
FIRST THREE STAGES: N_2O_4/UDMH
FOURTH STAGE: Block DM, liquid oxygen/kerosene; Breeze M, N_2O_4/UDMH

PERFORMANCE (PROTON M/ BREEZE M):
PAYLOAD TO LEO: 21,000 kg (46,300 lb)
PAYLOAD TO GTO: 6220 kg (13,710 lb)

PROPULSION:
FIRST STAGE: RD-253 (6)
SECOND STAGE: RD-0210 (4)
THIRD STAGE: RD-0212 (1)
FOURTH STAGE: Block DM, RD-58 (1); Breeze M, S5.98M (1)

FEATURES:
LENGTH: 57.6–58.2 m (188.9–190.9 ft)
CORE DIAMETER: 4.2 m (13.6 ft)
LAUNCH MASS: Proton K, 712,800 kg (1.6 million lb); Proton M, 690,000 kg (1.5 million lb)

Polar Satellite Launch Vehicle (PSLV) INDIA

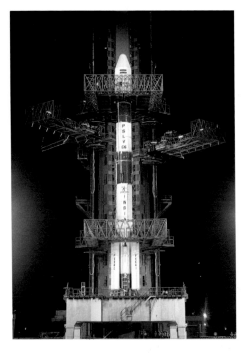

SPECIFICATION (PSLV C):

MANUFACTURER: ISRO
LAUNCH SITE: Sriharikota, India
FIRST LAUNCH: September 20, 1993

FUEL:
STRAP-ONS AND FIRST STAGE: Solid
SECOND STAGE: N_2O_4/UDMH
THIRD STAGE: Solid
FOURTH STAGE: N_2O_4/MMH

PERFORMANCE:
PAYLOAD TO LEO: 3700 kg (8100 lb)
PAYLOAD TO GTO: 1100 kg (2420 lb)

PROPULSION:
STRAP-ONS: S125 (6)
FIRST STAGE: S9 (1)
SECOND STAGE: Vikas (1)
THIRD STAGE: S7 (1)
FOURTH STAGE: L2 (2)

FEATURES:
LENGTH: 44.4 m (145.6 ft)
CORE DIAMETER: 2.8 m (9.1 ft)
LAUNCH MASS: 294,000 kg (648,000 lb)

The PSLV followed on from the SLV-3, a four-stage solid-propellant vehicle flown from 1979 to 1983, and the ASLV, which also had two strap-ons. It was developed to launch the Indian Remote Sensing (IRS) satellites into Sun-synchronous orbits, removing reliance on Russia. Since the first successful flight in October 1994, its Sun-synchronous orbit capability has been enhanced from 805 kg (1770 lb) to 1200 kg (2640 lb). PSLV can also launch a 3700 kg (8140 lb) satellite into LEO and a 1000 kg (2200 lb) satellite into GTO.

The operational "C" version, first flown in 1997, consists of two solid-propellant stages (1 and 3) and two liquid stages (2 and 4). The first stage is one of the largest solid-propellant boosters in the world and carries 138,000 kg (303,600 lb) of HTPB propellant. Six solid-fuel strap-on motors augment the first stage.

After some initial setbacks, the PSLV completed its first successful launch in 1996. It remains ISRO's primary launcher for LEO satellites. A modified version will be used to launch the Chandrayaan-1 lunar mission in 2008.

Rockot (SS-19) RUSSIA

SPECIFICATION:

MANUFACTURER: Khrunichev State
Research and Production Center
LAUNCH SITE: Plesetsk, Russia
FIRST LAUNCH: November 20,
1990

FUEL:
N_2O_4/UDMH

PERFORMANCE:
PAYLOAD TO LEO: 1950 kg
(4290 lb)

PROPULSION:
FIRST STAGE: RD-0233/
RD-0234 (4)
SECOND STAGE: RD-0235 (1)

FEATURES:
LENGTH: 29.2 m (95.8 ft)
DIAMETER: 2.5 m (8.2 ft)
LAUNCH MASS: 107,000 kg
(235,400 lb)

Rockot is a three-stage liquid-propellant launcher. Its first and second stages are based on the Russian SS-19 ICBM. During 1994 and 1995, Khrunichev and DaimlerChrysler Aerospace (now EADS Space Transportation) created a joint venture, Eurockot, which succeeded in booking a number of Western payloads.

Combined with the reignitable, highly maneuverable Breeze-KM upper stage, Rockot is capable of launching a 1950 kg (4290 lb) payload into LEO from Plesetsk. It is particularly suitable for launches of small and medium-sized spacecraft into Sun-synchronous, near-polar, and highly inclined orbits.

Rockot became available for commercial launches after a demonstration flight from Plesetsk in May 2000. In June 2003, Rockot launches from Plesetsk were temporarily banned because of the environmental threat from UDMH fuel. On October 8, 2005, a failure of the Breeze upper stage led to the destruction of ESA's CryoSat spacecraft. Rockot returned to flight on July 28, 2006, with the launch of South Korea's Kompsat-2.

Shavit (RSA-3, LK-A) ISRAEL

Shavit 1 pictured

SPECIFICATION
(Shavit 1):

MANUFACTURER: Israel Aircraft Industries
LAUNCH SITE: Palmachim, Israel
FIRST LAUNCH: September 19, 1988 (Shavit); April 5, 1995 (Shavit 1)

FUEL:
ALL STAGES: Solid (3 stages)

PERFORMANCE:
PAYLOAD TO LEO: 160 kg (350 lb)

PROPULSION:
FIRST STAGE: ATSM-13 (1)
SECOND STAGE: ATSM-9 (1)
THIRD STAGE: AUS-51 (1)

FEATURES:
LENGTH: 3.4 m (11 ft)
DIAMETER: 1.4 m (4.5 ft)
LAUNCH MASS: 27,250 kg (60,000 lb)

Shavit (Comet) is a solid-fuel launcher whose first two stages are derived from the Jericho II medium range ballistic missile. Developed to launch small satellites into LEO, the Shavit enabled Israel to become the eighth country to launch a satellite. Its first pay-load in September 1988 was the Israeli Ofeq 1 reconnaissance satellite.

The basic Shavit (RSA-3) launcher flew three times from 1988 to 1994. Its last launch was a failure. It was replaced by the Shavit 1 (LK-A), which has a larger, more powerful first stage. The latest failure of the Shavit in September 2004 resulted in the destruction of the Ofeq 6 spy satellite.

The Shavit 2 (LK-1)—similar to the Shavit 1, but with a larger second stage—was introduced with the launch of Ofeq 7 in June 2007. Also planned is the Shavit 3 (LK-2), which will use a Thiokol Castor 120 motor as its first stage.

It has been proposed to also launch the Shavit 2 from Wallops Island, Virginia, so that it can be launched eastward over the Atlantic. Another possibility is an air-launched version, using a standard Shavit 1 without a first stage that would be dropped from a Hercules C-130 aircraft.

Soyuz (SL-4, A-2)/Molniya (SL-6/A2e) RUSSIA

Soyuz-FG pictured

The Soyuz is best known as the launch vehicle for all crewed Soyuz spacecraft since 1967. It is also used to launch unmanned Progress supply spacecraft to the International Space Station and for commercial launches marketed and operated by the Russian–European Starsem company.

The Soyuz was derived from the R-7 ICBM and the Vostok launcher. It was initially a three-stage rocket, but in 1965 the four-stage Molniya (SL-6) variant was introduced, enabling it to reach the highly elliptical Molniya orbit or deliver payloads beyond Earth orbit. The addition of the restartable Ikar upper stage to the three-stage Soyuz in 1999 enabled 24 Globalstar satellites to be deployed in six launches.

The latest variant is the Soyuz 2-1B, which first flew on December 27, 2006. This vehicle has a new digital guidance system and a more powerful RD-124 third-stage engine that significantly increases the overall launch vehicle performance and payload mass capability. Launches of Soyuz ST (modified Soyuz 2-1A) vehicles from Kourou in French Guiana will begin in 2008.

SPECIFICATION:
(Soyuz U)

MANUFACTURER: TsSKB-Progress
LAUNCH SITES: Plesetsk, Russia; Baikonur, Kazakhstan;
FIRST LAUNCH: November 23, 1963 (Soyuz); May 18, 1973 (Soyuz U); December 27, 2006 (Soyuz 2-1B)

FUEL:

ALL STAGES: Liquid oxygen/kerosene

PERFORMANCE:

PAYLOAD TO LEO: 6855 kg (15,080 lb)

PROPULSION:

FIRST STAGE STRAP-ONS: RD-107 (4)
SECOND STAGE: RD-108 (1)
THIRD STAGE: RD-0110 (1)

FEATURES:

LENGTH: 45.2 m (148.3 ft); 49.5 m (162.4 ft) with Soyuz-TM escape tower
CORE DIAMETER: 3 m (9.7 ft)
LAUNCH MASS: 303,000 kg (666,600 lb)

Start-1 (SL-18, L-1, SS-25) RUSSIA

In the early 1990s, the Start launcher was developed by a group of Russian enterprises led by the MIHT Scientific and Technological Center. It is almost entirely derived from the Topol (SS-25) ICBM, which is the core of the Russian Strategic Rocket Forces. The original version made only one (unsuccessful) flight on March 28, 1995.

The Start-1, with a new fourth stage, consists of four solid-propellant stages and is launched from a mobile missile launcher on a seven-wheel axis truck. Up to the launch, the missile remains inside a container to protect it from mechanical damage and adverse weather conditions.

Start-1 can place a 550 kg (1210 lb) payload into LEO and payloads of 490–800 kg (1078–1760 lb) in near-polar circular orbits at altitudes of 200–1000 km (124–624 mi), with a wide range of inclinations. The first demonstration launch took place from Plesetsk in March 1993, with a 225 kg (495 lb) payload. There have been half a dozen commercial launches, most recently the Israeli EROS B imaging satellite on April 25, 2006.

SPECIFICATION:

MANUFACTURER: Moscow Institute of Heat Technology (MIHT)
LAUNCH SITES: Plesetsk, Russia; Svobodny, Russia
FIRST LAUNCH: March 25, 1993

FUEL:
ALL STAGES: Solid (4 stages)

PERFORMANCE:
PAYLOAD TO LEO: 800 kg (1760 lb)

PROPULSION:
FIRST STAGE: MIHT-1 (1)
SECOND STAGE: MIHT-2 (1)
THIRD STAGE: MIHT-3 (1)
FOURTH STAGE: Start-4 (1)

FEATURES:
LENGTH: 22.7 m (74.5 ft)
DIAMETER: 1.8 m (5.9 ft)
LAUNCH MASS: 47,000 kg (103,400 lb)

Strela (RS-18, SS-19) RUSSIA

SPECIFICATION:

MANUFACTURER: NPO Mashino-
stroyeniya
LAUNCH SITES: Svobodny, Russia;
Baikonur, Kazakhstan
FIRST LAUNCH: December 5, 2003

FUEL:
ALL STAGES: N_2O_4/UDMH
(2 stages)

PERFORMANCE:
PAYLOAD TO LEO: 1700 kg
(3750 lb)

PROPULSION:
FIRST STAGE: RD-0233 (4)
SECOND STAGE: RD-035 (1)

FEATURES:
LENGTH: 26.7 m (87.6 ft)
DIAMETER: 2.5 m (8.2 ft)
LAUNCH MASS: 104,000 kg
(229,000 lb)

Strela was converted by NPO Mashinostroyeniya into a launch vehicle from
the RS-18 ICBM's prototype, the UR-100N (SS-19) missile, with very few
changes apart from two fairing options and improved control-system
software. The two-stage Strela can be launched from an upgraded silo at
Svobodny or from Baikonur. A post-boost module is attached to the orbital
payload. Although Strela is slightly inferior to its Rockot (SS-19) competitor
in terms of payload mass, it has a lower orbital injection cost. It can put
satellites into orbits with an inclination of 63° from Baikonur and with
inclinations in the range 52–104° from Svobodny.

It was reported on December 11, 2002, that environmental concerns
prompted residents in the city of Blagoveshchensk to protest future launches
of Strela, which uses UDMH as a fuel source, from nearby Svobodny
Cosmodrome. The first Strela demonstration launch took place from
Baikonur in December 2003. A 978 kg (2150 lb) dummy spacecraft was
injected into a 458 km (284.6 mi) circular with an inclination of 67°.

Taurus USA

Taurus XL pictured

SPECIFICATION
(Taurus XL):

MANUFACTURER: Orbital Sciences Corporation (OSC)
LAUNCH SITE: Vandenberg, California
FIRST LAUNCH: March 13, 1994 (Taurus); May 20, 2004 (Taurus XL)

FUEL:
ALL STAGES: Solid (four stages)

PERFORMANCE:
PAYLOAD TO LEO: 1350 kg (3000 lb)
PAYLOAD TO GTO: 430 kg (950 lb)

PROPULSION:
FIRST STAGE: Castor 120 (1)
SECOND STAGE: Orion 50SXL (1)
THIRD STAGE: Orion 50XL (1)
FOURTH STAGE: Orion 38 (1)

FEATURES:
LENGTH: 27.9 m (91.5 ft)
DIAMETER: 2.4 m (7.7 ft)
LAUNCH MASS: 73,030 kg (161,000 lb)

The Taurus was developed by Orbital Sciences Corporation under a Defense Advanced Research Projects Agency (DARPA) contract for a demonstration launch of a "standard small launch vehicle." The four-stage all-solid rocket vehicle uses Pegasus upper stages with a Castor 120 first stage.

The first Taurus launch in 1994 delivered the USAF Space Test Experiment Platform and DARPASAT payloads to LEO. The second flight in February 1998 carried two Orbcomm satellites and a Celestis "burial" payload. However, a launch failure in September 2001 caused the vehicle to be grounded for almost three years. It returned to flight in May 2004 with the maiden launch of the uprated Taurus XL version, which uses the same upper stages as the Pegasus XL. This gives up to 25% more performance during launch.

OSC intends to offer the more-powerful Star 37 upper stage, which will increase performance to Sun-synchronous orbit from 945 kg (2080 lb) to about 1160 kg (2550 lb). Under a 2002 contract from Boeing, a three-stage version of Taurus was developed for use as interceptor boost vehicles for the US government's missile intercept system.

Tsyklon (R-36, SL-11, SL-14) UKRAINE

Tsyklon-3 pictured

SPECIFICATION:

MANUFACTURER: Yuzhnoye
Design Bureau
LAUNCH SITES: Baikonur,
Kazakhstan (Tsyklon-2); Plesetsk,
Russia (Tsyklon-3)
FIRST LAUNCH: Oct. 27, 1967
(Tsyklon-2); June 24, 1977
(Tsyklon-3)

FUEL:
N₂O₄/UDMH

PERFORMANCE:
PAYLOAD TO LEO: Tsyklon-2, 3200
kg (7040 lb); Tsyklon-3, 3600 kg
(7920 lb)

PROPULSION:
FIRST STAGE: Tsyklon-2, RD-251
(3); Tsyklon-3, RD-261 (3)
SECOND STAGE: Tsyklon-2,
RD 252 (1); Tsyklon-3, RD-262 (1)
THIRD STAGE: RD-861 (1)

FEATURES:
LENGTH: Tsyklon-2, 35.5–39.7 m
(116.4–130.2 ft); Tsyklon-3,
39.3 m (128.8 ft)
DIAMETER: 3 m (9.8 ft)
LAUNCH MASS: Tsyklon-2,
182,000 kg (400,400 lb);
Tsyklon-3, 186,000–190,000 kg
(409,200–418,000 lb)

The name Tsyklon (Cyclone) appeared for the first time in 1986 when the Soviets began promoting their launchers for commercial use in the West. The original Tsyklon launcher proposal, based on the Yangel Design Bureau's R-16 ICBM, never flew. However, the Tsyklon-2, based on the larger R-36 ICBM, did become operational in 1966. The two-stage rocket was initially used to launch large military payloads, including an antisatellite interceptor and ocean reconnaissance satellites.

Tsyklon-3, which featured a restartable third stage, was introduced in 1977. This version has been used for a wide variety of commercial, civil, and military payloads, often in multiple launches. The Tsyklon-2 (SL-11) is launched almost entirely from Baikonur, whereas the Tsyklon-3 (SL-14) is operated from Plesetsk. Production for both versions is now closed, but an updated version, known as Tsyklon-4, is currently under development. This includes modified lower stages, a new upper stage, and a 4 m (13.1 ft) diameter payload fairing. This version, which may be launched from Alcantara in Brazil, will be able to place a payload of 5250 kg (11,550 lb) into LEO.

Vega EUROPE

SPECIFICATION:

MANUFACTURER: ELV SpA
(Avio/ASI)
LAUNCH SITE: Kourou, French
Guiana
FIRST LAUNCH: 2008?

FUEL:
ALL STAGES: Solid (3 stages)

PERFORMANCE:
PAYLOAD TO LEO: 2000 kg
(4400 lb)

PROPULSION:
FIRST STAGE: P80 (1)
SECOND STAGE: Zefiro 23 (1)
THIRD STAGE: Zefiro 9 (1)

FEATURES:
LENGTH: 30.2 m (99.1 ft)
DIAMETER: 3 m (9.8 ft)
LAUNCH MASS: 137,000 kg
(301,400 lb)

Vega has been under development since 1998, with the support of seven ESA member states (Italy, France, Belgium, Switzerland, Spain, the Netherlands, and Sweden). The small satellite launcher is an all-solid three-stage vehicle with a liquid-fuelled injection module used for attitude and orbit control and satellite release. ELV SpA, a joint venture of Avio and ASI, the Italian space agency, is responsible for Vega development. CNES, the French space agency, holds similar responsibility for the P80 first stage.

Vega is designed to loft single or multiple payloads to orbits up to 1500 km (932 mi) in altitude. Its baseline payload capability is about 1500 kg (3300 lb) to a circular 700 km (435 mi) high Sun-synchronous orbit, but it can also loft satellites from 300 kg (660 lb) to more than 2000 kg (4400 lb), as well as piggyback microsatellites. Once qualified, Vega will be marketed and operated by Arianespace for the small- to mid-sized satellite launch market. The first launch is planned for 2008 from Europe's Kourou spaceport in French Guiana, where the Ariane 1 launch facilities have been adapted for its use.

Volna (R-29RL, RSM-50, SS-N-18) RUSSIA

SPECIFICATION:

MANUFACTURER: Makeyev Design Bureau
LAUNCH SITE: Nuclear missile submarine (Barents Sea)
FIRST LAUNCH: June 6, 1995

FUEL:
N₂O₄/UDMH

PERFORMANCE:
PAYLOAD TO LEO: 120 kg (264 lb)

PROPULSION:
FIRST STAGE: RD-0243 (1)
SECOND STAGE: ? (1)
THIRD STAGE: ? (1)

FEATURES:
LENGTH: 14.1 m (46.3 ft)
DIAMETER: 1.8 m (5.9 ft)
LAUNCH MASS: 35,200 kg (77,440 lb)

The Volna (Wave) is a R-29L submarine-launched ballistic missile (called SS-N-18 by NATO) built by the Makeyev company. Approximately 350 missiles were built after it entered service in 1977. The submarine cruises for 4 or 5 h to its designated launch station.

The basic three-stage Volna is only capable of suborbital missions, but a Volna-O version with a small liquid-fueled fourth stage developed by Babakin can launch small payloads to LEO. The standard Volna commercial microgravity orbit is about 2200 x 200 km (1367 x 124 mi) with an inclination of 76°.

Reliability is an issue for the Volna. Three of five commercial launches have been unsuccessful, with two of these failures—both involving the Cosmos-1 solar sail—due to a malfunction of the missile.

Other Volna missions have carried the IRDT inflatable reentry device, designed to return cargo from orbit to Earth. After third-stage separation, the IRDT fires a boost motor to increase its speed and then inflates the first stage of its heat shield.

Zenit (Zenith, SL-16, J-1) UKRAINE/RUSSIA

Zenit-3SL pictured

SPECIFICATION:

MANUFACTURER: SDO Yuzhnoye/
PO Yuzhmash
LAUNCH SITES: Baikonur,
Kazakhstan (Zenit-2, Zenit-2SLB,
and Zenit-3SLB); Sea Launch
platform, Pacific Ocean (Zenit-
2SL and 3SL)
FIRST LAUNCH: April 13, 1985
(Zenit-2); March 28, 1999
(Zenit-3SL)

FUEL:
Liquid oxygen/kerosene

PERFORMANCE:
PAYLOAD TO LEO: Zenit-2,
13,740 kg (30,290 lb)
PAYLOAD TO GTO: Zenit-3SL,
6160 kg (13,552 lb)

PROPULSION:
FIRST STAGE: RD-171 (1)
SECOND STAGE: RD-120 (1)
THIRD STAGE: RD-58M (1)

FEATURES:
LENGTH: 57.4–58.7 m
(188.3–192.4 ft)
DIAMETER: 3.9 m (12.8 ft)
LAUNCH MASS: Zenit-2,
445,204 kg (981,497 lb);
Zenit-3SL, 535,998 kg
(1.2 million lb)

The Zenit was developed in the 1980s for two purposes: as a liquid strap-on booster for the Energia rocket and, equipped with a second stage, as a launcher in its own right. At that time, it was expected to replace the Soyuz launcher for manned missions, but these plans were abandoned after the fall of the Soviet Union. Four Zenit strap-on boosters were successfully flown on both flights of Energia.

The Zenit-2 has been launched from Baikonur Cosmodrome in Kazakhstan since 1985. A three-stage version, the Zenit-3SL, has been used for commercial launches from the Sea Launch consortium's floating launch platform in the Pacific Ocean since 1999.

The first- and second-stage engines, as well as the Block-DM upper stage of the Zenit-3SL, are supplied by Russian companies Energia and Energomash. A commercial launch service from Baikonur is expected to begin in 2008, using a Zenit-3SLB rocket equipped with a modified Block-DM third stage manufactured in Russia by RSC Energia. The Zenit-M, with a modified first-stage engine and flight-control system, first flew on June 29, 2007.

Launch Sites

Alaska Spaceport (Kodiak Launch Complex) USA

57.43°N, 152.33°W

The Alaska State Legislature created the Alaska Aerospace Development Corporation (AADC) in July 1991 to promote development of the aerospace industry—including space launch services—in the state. In 1995, the state legislature approved funding for the Kodiak Launch Complex. In 1996, the USAF Atmospheric Interceptor Test Program committed to two launches from Kodiak. Bids for construction were solicited in 1997 and construction began in 1998. That same year, AADC signed its first contract with a commercial customer, Lockheed Martin Corporation. Construction was completed in 2000.

Alaska Spaceport covers an area of 15 km^2 (5.8 mi^2)and is located on Narrow Cape, Kodiak Island. The hilly, almost-treeless island is a volcanic peak in the ocean, 48.2 km (30 mi) off the southern coast of Alaska. It is about 400 km (248.5 mi) south of Anchorage and 40 km (24.8 mi) southwest of the city of Kodiak. The climate is more moderate than its northerly latitude would suggest, with a yearly mean temperature of 4°C (39.2°F) and only three months in the year where the average temperatures fall below zero (32°F). Visibility and prevailing winds compare favorably with those at Vandenberg Air Force Base in California.

The site is promoted as one of the best locations in the world for polar launches, providing a wide launch azimuth and unobstructed downrange flight path over the North Pacific Ocean, avoiding populated areas.

Launch Pad 1, used for launching orbital missions, is serviced by an enclosed and movable gantry called the Launch Service Structure (LSS). Launch Pad 2, used for suborbital launches, is enclosed and serviced by the Spacecraft and Assemblies Transfer (SCAT) facility. The SCAT, a movable structure roller-mounted on rails, also provides all-weather indoor transfer of processed motors into the LSS. Other facilities include the Launch Control Center, a Payload Processing Facility, and an Integration Processing Facility.

The only orbital launch was by an Athena-I rocket carrying the Kodiak Star payload of four NASA and DOD satellites on September 29, 2001. In 2003, AADC and the US Missile Defense Agency entered into a five-year contract for numerous suborbital launches in connection with tests of the nation's missile defense system. The first launch under this contract took place on December 14, 2004. The most recent launch, on February 23, 2006, involved the launch of the Missile Defense Agency's Flight Test rocket FT 04-1 in support of a target missile tracking test. The spaceport also offers backup to Vandenberg Air Force Base for satellites needing delivery to polar orbit.

Alcântara BRAZIL

2.28°S, 44.38°W

The Centro de Lançamento de Alcântara (CLA) is a satellite launching base near the city of Alcântara, located on Brazil's North Atlantic coast, in the state of Maranhão. It is operated by the Brazilian Air Force (Comando da Aeronáutica). The CLA is closer to the equator than any launch site in the world. This gives it a significant advantage in launching payloads such as geosynchronous satellites towards the east: launches gain 25% in launch energy due to Earth rotation compared with those from Cape Canaveral. The coastal location also gives a wide launch azimuth to the north and east over the ocean.

Construction of the 520 km^2 (200.7 mi^2) base began in 1982. Brazilian government and military officials held a formal opening ceremony in February 1990; on February 21, the facility had its first launch—the Sonda 2 XV-53 sounding rocket. In addition to Brazilian sounding rockets, meteorological rockets, and other suborbital scientific flights, the Ongoron I and Ongoron II rockets were tested here by the French government, and NASA flew four Nike Orion vehicles in 1994.

The launch site has been expanded for operations of the VLS orbital launcher. Unfortunately, none of the VLS launches have been successful. On August 22, 2003, the explosion of the third VLS-1 rocket (VO3) killed 21 people and destroyed the launch pad.

There are also plans to launch several international rockets from Alcântara. In 2003, contracts were signed to launch Ukrainian Tysklon-4 and Israeli Shavit rockets. There are also tentative plans to launch the Russian Proton and the Chinese Long March 4.

Current government plans to construct a civilian launching center, operated by the Brazilian Space Agency (Agência Espacial Brasileira) and adjacent to the military-controlled CLA, has stirred opposition among the local population, who are concerned about relocation.

Baikonur, Tyuratam KAZAKHSTAN

45.6°N, 63.4°E

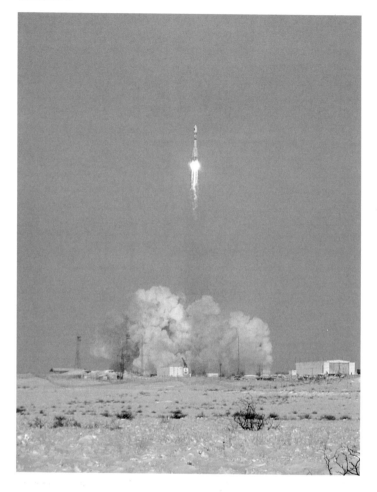

Russia's largest cosmodrome is located on the flat steppes near the town of Tyuratam in Kazakhstan. It has an extreme climate, with bitterly cold winters and hot summers. In the cold war, it was misleadingly given the name Baikonur by the Soviet authorities in order to hide its precise location. Today, the spaceport is leased from the Kazakh government. In 2005, Kazakhstan agreed to continue leasing Baikonur to the Russian Federation until 2050.

The cosmodrome extends for 85 km (53 mi) north to south, and 125 km (78 mi) east to west. Aside from dozens of launch pads, it includes five tracking-control centers, nine tracking stations, and a 1500 km (930 mi) rocket test range. The cosmodrome has traditionally been split into three sectors: the Right and Left flanks and the Center. Nine launch complexes with a total of 15 pads are available. It is the only cosmodrome supporting Proton and Zenit launches. It also provides services for Soyuz, Molniya, Tsyklon, Rockot, Kosmos, and various missile and missile-derived rocket launches. Future Angara launches are expected to take place from there.

The initial R-7 launch complex, Area 1, in the Center sector, was built for missile and rocket tests, which began in 1955. The first pad built at Baikonur launched both Sputnik 1 and Yuri Gagarin into orbit. Today, the refurbished pad hosts Soyuz launches. Azimuth launches from due east (the most efficient) are prohibited because the lower stages would impact in China.

The Left Flank is mainly occupied by the launch pads, assembly buildings, and housing for the Chelomei design bureau. It includes launch pads for the Tsyklon-2 and Proton. In the Centre Flank are facilities for the projects of the Korolev OKB-1 design bureau, including the R-7/Vostok/Soyuz launch pad and the N-1 Moon program facilities (later converted for use in the Energia–Buran program). The Right Flank primarily houses facilities for Yangel's design bureau. It includes a second R-7 pad and a Zenit launch pad.

Baikonur town, previously known as Leninsk, was founded May 5, 1955, as the cosmodrome's residential area. It has an official population of about 60,000, down from a peak of 100,000 in the mid-1980s. The cosmodrome has several hotels and residential areas, closer to launch facilities and separate from the main town.

Today, all Russian manned flights and planetary missions are launched from Baikonur. Until now, it has been operated by Russian Space Forces, but they are about to hand over to the Russian Federal Space Agency.

Cape Canaveral Air Force Station, Florida USA

28.5°N, 81.0°W

Cape Canaveral Air Force Station lies on a sandy peninsula on the Florida coast that is also a nature reserve. It originated as a missile test center on the site of an old air base. The construction of permanent facilities began in 1950, when it became known as the Cape Canaveral Air Force Station at Patrick Air Force Base. It is part of the US Air Force Command 45th Space Wing. From 1964 to 1973, Cape Canaveral was known as Cape Kennedy.

The station's Eastern Range tracking network extends all the way into the Indian Ocean where it meets the Western Range network. Launches take place toward the east (over the Atlantic Ocean) with orbital inclinations up to 57° Polar launches are not permitted because they would have to fly over populated areas. The first launch from the test center took place on July 24, 1950. It was a Bumper rocket, a V-2 missile carrying a WAC Corporal second stage.

With the launch of Explorer 1 by a Jupiter-C rocket on January 31, 1958, the station became the main center for US civil launches. Three years later, with the beginning of the Mercury program, it became the only site for US manned launches. Since then, more than 500 space launches have been made from the station, including NASA's manned missions. The annual launch rate reached 25–30 flights during the 1960s. There were eight space launches (including three shuttle flights) during 2006.

The 45th Space Wing provides extensive support for Space Shuttle launch operations. The USAF Eastern Range provides weather forecasting, range safety, tracking, and preparation of DOD payloads. For over four decades, launches took place from 47 different complexes; today, most launch activities are handled by just a few. Space Launch Complex (SLC) 41 launches the Atlas V, SLC 37 launches the Delta IV, and SLC 17 launches the Delta II. (SLC 41 is actually north of the Air Station boundary in Kennedy Space Center, on a site allocated to the USAF.)

Launch Complex 39, located on Merritt Island to the north of the Cape Canaveral base, was added in the mid-1960s for Saturn V launches and administered by NASA. This area is known as the John F. Kennedy Space Center and now has two launch pads for the Space Shuttle. There is also a commercial launch site at Cape Canaveral Air Force Station operated by the Spaceport Florida Authority. It has converted the Navy's old SLC 46 pad for launching small to medium commercial launchers to equatorial orbit. Core rocket stages for large vehicles such as the Delta IV are delivered by special cargo ships to nearby Port Canaveral.

Jiuquan (Shuang Cheng Tzu) CHINA

40.6°N, 99.9°E

Jiuquan Satellite Launch Center was built north of Jiuquan City in the Gobi desert of Inner Mongolia, 1600 km (1000 mi) west of Beijing. The site is known in the West as Shuang Cheng Tzu. The site is about 1000 m (3300 ft) above sea level. It has long, cold winters and hot summers. Annual rainfall is

40 mm (0.2 in), but Jiuquan lies close to the Ruoshui River. This has made it possible to irrigate the area, with the development of more than 60 oases. To store water, a man-made reservoir covering 10 km^2 (3.2 mi^2) was built in the town. The site comprises three separate zones—the launch zone, the red willow and poplar zone, and the urban zone. Construction of China's first and largest launch site began in 1958.

The first launch, involving a Russian-built R-2 vehicle, took place on September 1, 1960. Jiuquan was also the location for the first orbital space launch by the People's Republic of China when a Long March 1 rocket launched the Mao 1 satellite on April 24, 1970. China then became the fifth nation to launch an artificial satellite into orbit.

Today, Jiuquan is limited to southeastern launches into 57–70° orbits to avoid overflying Russia and Mongolia. There are two launch pads 416 m (1365 ft) apart, with a shared mobile service tower. Road and rail systems for transport of satellites and rocket stages link the site to an airport 94 km (151 km) to the south. Rockets of the Long March 1 and 2 series are launched from Jiuquan, carrying recoverable Earth observation and microgravity payloads. Jiuquan is also the launch center for the Long March 2F, which carries China's manned Shenzhou spacecraft. Due to the site's northerly geographical location, most Chinese commercial flights take off from other spaceports.

Northeast of the launch base is the burial site of more than 500 people who contributed to the country's space cause, including the founder of China's space program, Marshal Nie Rongzhen.

Kapustin Yar RUSSIA/KAZAKHSTAN

48.5°N, 45.8°E

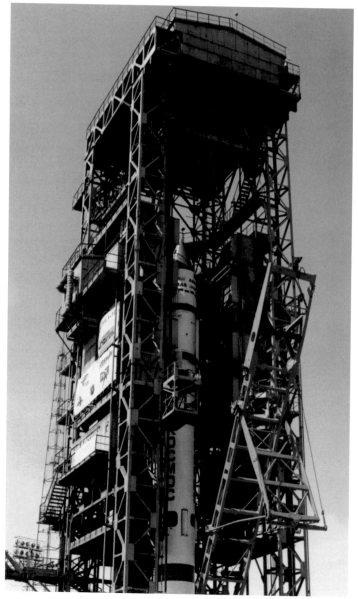

Kapustin Yar lies near the Volga river, partly in the Volgograd and Astrakhan regions of Russia and partly in western Kazakhstan. It has also been called the Volgograd Station, the Fourth State Central Test Site, and the State Central Inter-Service Test Site. It occupies a total of 65,000 km² (25,097 mi²) of desert. Following the break-up of the USSR, the Kazakhstan portion of the site was leased to Russia.

First established by government decree in 1946, it was the first test range for Soviet missiles and sounding rockets, with the first launch of a modified V-2 missile on October 18, 1947. It was greatly expanded as the test site for numerous Soviet intermediate and short-range missile projects in the 1950s. Kapustin Yar was also the headquarters of the first operational R-1/R-2 units from 1950 to 1953, and later a base for 12 operational R-14 missile launchers. All launches are towards the east.

A new town, Znamensk, was established for the scientists, supporting personnel, and families working on the site. For many years, the Soviets tried to keep its existence a secret. Evidence of its importance was obtained by Western intelligence through debriefing of returning German scientists and spy flights.

The first orbital launch from Kapustin Yar was Kosmos 1 in 1962. By 1965, two silo complexes were converted for the Kosmos 2 launcher. Since 1973, the Kosmos 3M launcher has flown from this base. By 1980, 70 space launches to orbit were carried out, mostly small Kosmos science satellites. The USSR then concentrated its classified space launches at Plesetsk, leaving Kapustin Yar to send up only occasional missions, usually for radar calibration. Between 1980 and 1988, a modified version of the Kosmos 3M rocket launched 10 scaled prototypes of a reusable shuttle to test the aerodynamics of the future Buran orbiter.

There has only been one orbital launch since 1987, although there have been some missile-testing activities as well as Kosmos suborbital launches. The last orbital Kosmos 3M launch from Kapustin Yar, a commercial flight for Germany and Italy, took place on April 28, 1999, after an 11-year break.

Kourou (Centre Spatial Guyanais) FRENCH GUIANA

5.2°N, 52.8°W

The site at Kourou, in the colony of French Guiana, South America, was selected in April 1964 after Algeria gained independence from France and access to launch sites in that country was lost. With its location near the equator and near the Atlantic coast, Kourou is one of the best launch sites in the world. The near-equatorial location allows maximum energy assist from the Earth's rotation for launches into equatorial orbits, giving an extra 460 m/s (1500 ft/s) in velocity. Weather conditions are favorable throughout the year. The shape of the coastline allows launches between north and east (-10.5° to +93.5°)

Although the site is owned by the French space agency, CNES, it has been used as the main European spaceport since July 1966, when ELDO chose the site for launches of the Europa II launch vehicle. Today, it is used by the ESA and the commercial launch company Arianespace.

Four pads for sounding rockets were completed in 1968 and a Diamant pad in 1969. The Europa II launch complex was ready in 1971 but was used for only one launch before the project was terminated. At the time of the last

Diamant orbital launch in 1975, 184 sounding rocket and nine orbital launches had been made from Kourou.

The Europa II launch complex (ELA-1) was modified for use with the Ariane, and since December 1979, Kourou has been the launch center for all Ariane flights. A second pad, ELA-2, was completed in 1986 and used for Ariane 4 launches. The introduction of Ariane 5 led to the addition of ELA-3, completed in 1996. After the final Ariane 4 launch in 2003, ELA-1 and ELA-2 were decommissioned. However, ELA-1 is now being refurbished for use by the Vega launcher, with a first flight planned in 2008. Launcher assembly and integration will be performed on the pad within a new mobile gantry. A new pad, 10 km (6 mi) north of ELA-1, is being built for use by the Russian Soyuz 2 (Soyuz ST) launcher, with the maiden flight planned for early 2009. Rocket stages are delivered by sea to a nearby port, while most payloads arrive by air.

Kwajalein MARSHALL ISLANDS

9.28°N, 167.73°E

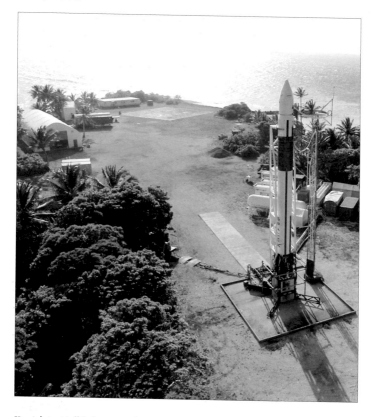

Kwajalein Atoll belongs to the Republic of the Marshall Islands in the South Pacific. Located about 3380 km (2100 mi) southwest of Hawaii and 2250 km (1400 mi) east of Guam, it is the home to the only US equatorial launch facilities. These facilities are known as the Ronald Reagan Ballistic Missile Defense Test Site and operated by the US Army under a long-term lease. The Reagan Test Site encompasses approximately 1.9 million km² (750,000 mi²), although the total land area is only about 180 km² (70 mi²). It includes rocket launch sites on 11 islands in the Kwajalein Atoll, on Wake Island, and at Aur Atoll. About 2000 support personnel and family members live on Kwajalein and Roi-Namur islands.

One of the launch sites is on a tiny island called Omelek. The island is part of a coral reef and covers about 32,000 m² (8 acres). The island's proximity to the equator allows additional payload to be launched to the east, and space launches require a smaller plane change when launched from Kwajalein rather than from other sites. However, the site can accommodate launches for almost any orbital inclination. Due to its relative isolation, Omelek has long been used by the United States military for launches of small research rockets. The last US government launch occurred in 1996. More recently, a new commercial launch company, SpaceX, updated facilities on the island for the first flights of their Falcon 1 rocket. The first launch attempt of the Falcon 1 in March 2006 was cut short by an engine fire; the second test flight on March 20, 2007, was more successful.

Launch vehicles arrive at Kwajalein via commercial cargo carrier and are transferred to Omelek by landing craft. Four days before launch, a trailer carries the launcher, which is fitted to the erector, to the pad in a horizontal position. The island is evacuated during the launch. The rocket's first stage is fitted with a parachute system that enables it to fall into the sea and be recovered by ship for possible refurbishment and reuse.

Odyssey (Sea Launch) USA

0°N, 154°W

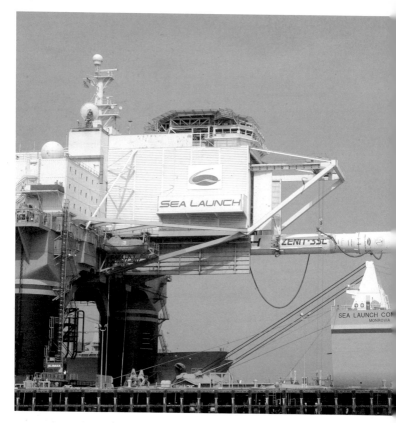

The Odyssey launch platform is operated by Sea Launch, an international company of American, Russian, Ukrainian, and Norwegian partners. Starting up in 1995, the company's first launch took place on March 27, 1999, when a Zenit-3SL vehicle completed a demonstration launch of a test payload.

Launch operations begin at Home Port in Long Beach, California, where the satellite is delivered. Following the fueling and encapsulation of the satellite in the payload processing facility, the integrated payload unit is transferred to the Sea Launch Commander Assembly and Command Ship (ACS) for integration with the launch vehicle. Then, the horizontally integrated rocket is transferred to the launch platform, where it is stored for transit to the equator.

Two unique ships form the Sea Launch marine infrastructure. The ACS is a specially designed vessel that serves as a floating rocket-assembly factory while in port; provides accommodation and leisure facilities for up to 240 crew members, customers, and VIPs; and houses mission-control facilities for launches at sea. The ACS is 200 m (660 ft) long and approximately 32.3 m (106 ft) wide, with a displacement of more than 34,000 t (68 million lb). It has a cruising range of 33,336 km (18,000 nautical mi).

The launch platform is a former North Sea oil-drilling platform. Refurbished in Stavanger, Norway, the vessel is one of the largest semi-submersible, self-propelled vessels in the world. It measures 133 m (436 ft) long and about 67 m (220 ft) wide, with an empty draft displacement of 30,000 t (60 million lb). The platform provides accommo-dation for 68 crew and launch-system personnel. It has a large, environmentally controlled hangar for storage of the Zenit rocket and mobile transporter/erector equipment that is used to roll out and raise the rocket prior to fueling and launch. It also stores the rocket fuel (kerosene and liquid oxygen) for each mission.

The launch platform takes 11–12 days to reach the equatorial launch site in the mid-Pacific. It is then ballasted to its launch depth and turned to a specific heading to minimize wind and wave effects. Countdown, launch, telemetry, and tracking are dealt with from the launch control center on the Sea Launch Commander by English- and Russian-speaking launch teams. The rocket is rolled out of its environmentally protected hangar and automatically erected on the launch pad 27 hours prior to liftoff. Launch-support personnel transfer to the ship prior to propellant loading.

The platform was damaged during a launch failure on January 30, 2007, but operations should resume in January 2008 (at the earliest).

Plesetsk, Mirnyy RUSSIA

62.7°N, 40.3°E

The Plesetsk launch site is located close to the Arctic Circle, 170 km (105 mi) south of Archangel at Mirnyy in Arkhangelsk Oblast and about 800 km (500 mi) north east of Moscow. It is situated south of the river Emtsa in a relatively flat taiga (coniferous forest) landscape among numerous lakes. Winters are cold, down to -38°C, with a fair amount of snow, while summers are quite hot, up to 33°C.

Established on July 15, 1957, as the secret "Angara" facility, Plesetsk has been a leading rocket-testing and space-launch complex for over 40 years. Its existence was only officially acknowledged by the Soviets in 1983. The launch site, which is operated by the Russian Ministry of Defense, covers 149 km² (575.3 mi²) and includes 1745 experimental-technical facilities. About 40,000 service personnel and their families live in the nearby town of Mirnyy.

Despite its importance as a military base, Plesetsk suffered from financial difficulties after the fall of the Soviet Union, and in September 1995, its electricity was cut off for three days due to a failure to pay its power bills. There have been at least three launch-pad disasters in the history of Plesetsk: those in 1973 and 1980 cost a total of 60 lives. Most recently, a launch explosion in October 2002 killed one man, injured eight people, and severely damaged the Soyuz launch pad.

Plesetsk was a originally a missile defense facility where R-7 intercontinental missiles were based. In 1964, some of the facilities were converted for civilian space launches. Its first satellite launch took place on March 17, 1966, when a Vostok-2 booster placed the secret Kosmos-112 spacecraft into orbit. Since that time, more than 1500 missiles and rockets have been launched from Plesetsk—more than from any other launch site in the world.

There are currently eight operational launch pads, with facilities for Kosmos 3M, Soyuz-U/ Molniya-M, Rockot, Start-1, and Tsyklon-3 vehicles. A launch pad was being built for planned Zenit launches but these were dropped, and the partly finished site is now set aside for the Angara. The rockets are delivered by rail from the assembly and test facilities in a horizontal position and then erected on the pad. Range restrictions now limit launches to between 62.8° and 83° for polar and high-inclination orbits.

Spaceport America (Southwest Regional Spaceport) USA

33°N, 107°W

Spaceport America, formerly known as Southwest Regional Spaceport, was conceived by the Southwest Space Task Force, a private group of space activists who were keen to develop a commercial launch site in New Mexico that would also be suitable for reusable launch vehicles. Based on years of study, they selected 70 km² (27 mi²) of state-owned land, 72 km (45 mi) north of Las Cruces. This area of flat, open range to the west of White Sands Missile Range lies 1389 m (4557 ft) above sea level.

In 2003, the governor and state legislature accepted the idea of developing the site as America's premiere inland commercial spaceport. A year later, New Mexico hosted the X Prize Cup annual spaceflight exhibition. On December 13, 2005, it was announced that Virgin Galactic was to undertake a joint venture with New Mexico to construct a $225 million facility to be known as Spaceport America at Upham. The venture was approved by the state legislature and the funding plans signed into law on March 1, 2006.

By the end of 2006, the list of spaceport tenants also included UP Aerospace, Starchaser Industries, and the Rocket Racing League. UP Aerospace made the first (suborbital) rocket launch from temporary facilities at the Spaceport on September 25, 2006, using the SpaceLoft XL solid-fuel rocket. Although this was a failure, a second suborbital flight on April 28, 2007, was successful. The spaceport is expected to become operational by late 2009 or early 2010. Studies project that 2300 people will be employed at Spaceport America by its fifth year of operation. Future infrastructure will include a launch complex; a 3600 m (11,800 ft) long runway suitable for handling White Knight 2 tourist craft returning from suborbital flights conducted by Virgin Galactic; a payload assembly complex; and a futuristic terminal.

Sriharikota (Satish Dhawan Space Center) INDIA

13.7°N, 80.2°E

Sriharikota, which is operated by the Indian Space Research Organization (ISRO), is the main Indian launch center. It is located on Sriharikota Island on the east coast of Andhra Pradesh, southern India, about 100 km (66 mi) north of Chennai. It was previously called Sriharikota Range and Sriharikota Launching Range. It was given its current name in 2002, after the death of ISRO's former chairman, Satish Dhawan.

The center became operational in October 1971, with the launch of three Rohini sounding rockets on October 9 and 10. On July 18, 1980, India became the eighth nation to launch an artificial satellite, launching Rohini 1, using an SLV (Satellite Launch Vehicle). Today, in addition to sounding-rocket facilities, the center has two launch pads for orbital vehicles. The PSLV launch complex was commissioned in 1990. It has a 3000 tonne (6.6 million lb), 76.5 m (250 ft) high mobile service tower, which provides the payload clean room. The tower is rolled back for launch.

The second complex was introduced in May 2005 as a universal launch pad, able to accommodate both the PSLV and the larger, more powerful GSLV. The two launch pads allow more-frequent launches. Sriharikota's proximity to the equator is especially advantageous for launching payloads eastward, although safety considerations limit these to azimuths less than 140°.

The center also has facilities for solid-propellant processing; static testing of solid motors; launch-vehicle integration and launch operations; range operations comprising telemetry, tracking, and command network; and a mission control center. Approximately 2400 people are employed at the center.

Svobodny RUSSIA
51.4°N, 128.3°E

Svobodny was established in 1968 as a secret ICBM base for strategic missile forces. It was closed in 1993 after an agreement on the Strategic Arms Reduction Treaty II (START II). It was reactivated as a cosmodrome by President Boris Yeltsin on March 1, 1996, during negotiations with Kazakhstan over the future of Baikonur. The plans called for the restoration of up to five launch silos for UR-100-type missiles, which could be used for the Rockot and Strela converted missiles.

The cosmodrome is located at Svobodny-18, about 97 km (60 mi) from the Chinese border. Although the division of the Strategic Rocket Forces stationed in Svobodny-18 has been disbanded, the town still has a population of 6000. Despite its remote location and extreme climate, Svobodny has the advantage of being closer to the equator than Plesetsk, with the result that a launcher can deliver a payload up to 25% heavier than would be possible from Plesetsk.

Svobodny is available for launches of both military and commercial satellites. Five satellites have been launched since its reactivation in 1996. The first of these took place on December 24, 1997, when the Zeya experimental satellite was placed into orbit. The most recent launch was on April 25, 2006, carrying the Israeli Eros-B1 remote sensing satellite. All of the launches used Start-1 boosters fired from mobile launchers.

Future launches have been proposed for Strela, Rockot, and Angara boosters, but the necessary construction was delayed due to lack of funds. In early 2007, it was reported that the seldom-used launch site will be shut down again in the near future. Only a military unit and a ground measurement station would remain at the site.

Taiyuan (Wuzhai Missile and Space Test Center) CHINA

37.5°N, 112.6°E

Taiyuan Satellite Launch Center is located in Kelan County, northern Shanxi Province, nearly 300 km (480 mi) away from Taiyuan city, the provincial capital, and 500 km (800 mi) southwest of Beijing. The climate is fairly dry (540 mm or 21 in of rain per year), with cold winters and fairly hot summers. Hemmed in by mountains to the east, south and north, and the Hwang Ho (Yellow River) to the west, the center stands at an elevation of 1500 m (4920 ft) in sparsely populated terrain. US Space Command refers to the site as Wuzhai Missile and Space Test Center.

The formerly secret center was founded in March 1966 and began operation as a missile testing range in 1968. The first photographs of the site were not released until 1990. Today, it is also used as a launch center for government and commercial satellites. It has a single launch pad, opened in 1988, for launching Long March rockets.

The first Long March 4 was launched from Taiyuan on September 6, 1988. Since then, the Long March 4 has been used to carry all of the Chinese Sun-synchronous meteorological satellites, as well as remote sensing and reconnaissance satellites to polar and Sun-synchronous orbits.

Long March 2C rockets carrying US Iridium satellites were also launched from Taiyuan in the 1990s. The second Double Star satellite, an ESA–Chinese joint project designed to study the Earth's magnetosphere, was launched from Taiyuan in 2004.

Taiyuan is equipped with a command and control center as well as tracking and telemetry facilities. It is served by two railways that connect with the Ningwu–Kelan railway.

Tanegashima JAPAN

30.4°N, 131.0°E

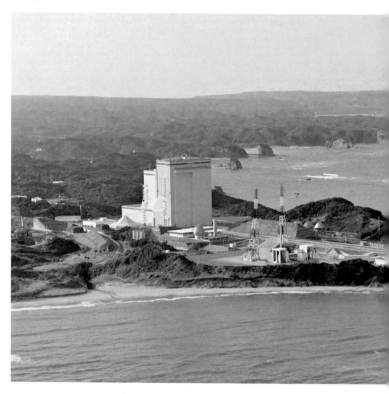

The Tanegashima Space Center is on the south eastern tip of Tanegashima Island, 50 km (31.1 mi) off the southern coast of Kyushu and 1050 km (650 mi) south west of Tokyo. It is the largest rocket range in Japan, with a total area of 8.6 km² (3.3 mi²). The landscape of blue sea, cliffs, and white sandy beach has led to its being called "the most beautiful launch site in the world." Formerly the main launch center for NASDA (see JAXA entry), it is now operated by Japan's space agency, JAXA.

There are two main facilities. The Osaki Range in the north includes the Yoshinobu Launch Complex, from which H-IIA rockets are launched, and a second pad from which the J-I rocket was launched. This includes the Vehicle Assembly Building, the Spacecraft Test and Assembly Building, and the Block House for countdown operations. It also has static test facilities for liquid-fuel rocket engines. The Takesaki Range to the south handles sounding

rockets and provides facilities for H-II solid-booster static firings and the H-II Range Control Center. With few exceptions, departures from both ranges are normally restricted to two launch seasons (January 15–end of February, and August 1–September 15) because of range safety procedures and agreements with the fishing industry.

The first orbital launch from Tanegashima took place on September 9, 1975, with a N-I rocket and its Kiku-1 satellite. On February 23, 1977, it became the third site in the world from which geostationary satellites could be launched. The N-I, N-II, H-I, and J-1 have all flown from Tanegashima. Today, the sole orbital launcher is the H-IIA, which is launched from a site 1 km (0.6 mi) away from the old H-I pad, which closed in 1992. By the end of 2006, the center had conducted 42 orbital launches.

Uchinoura, Kagoshima JAPAN

31.2°N, 131.1°E

Uchinoura Space Center is the launch site for solid-fuel sounding rockets and satellite launchers developed and launched by Japan's Institute of Space and Aeronautical Science (ISAS), now part of JAXA. It is located in a hilly region on the east coast of Ohsumi Peninsula, Kagoshima Prefecture, on the southern tip of Kyushu Island. The center covers an area of 0.7 km² (0.3 mi²).

Facilities for launching rockets, telemetry, tracking, and command stations for rockets and satellites, and optical observation posts have been built by flattening the tops of several hills. Antennas 20 m (66 ft) and 30 m (98 ft) in height receive telemetry from satellites to track and control them while in Earth orbit. The 34 m (112 ft) antenna can back up the 64 m (210 ft) antenna at the Usuda Deep Space Center.

Following the closure of the original facility Iwaki (now Yuri-Honjo) Akita Prefecture, the center was founded in February 1962, as part of the Institute of Industrial Science, University of Tokyo. In 1964, it became part of the University's Institute of Space and Aeronautical Science. Seventeen years later, the Kagoshima Space Center became an independent research facility, part of ISAS. It was renamed the Uchinoura Space Center on October 1, 2003, aftre the creation of JAXA.

Launches are limited to two months each year in order to avoid disturbing local fisheries. Despite this, more than 360 rockets and 23 satellites and probes were launched from the center between 1970 and 2003. The site was originally used for atmospheric sounding rockets and meteorological rockets. The first orbital flight took place on February 11, 1970, with the launch of the Ohsumi satellite by a Lambda 4S-5 rocket, making Japan the fourth nation to launch an artificial satellite. The largest launch vehicle operated from Uchinoura was the M-V, first launched in 1997. The most recent launch from Uchinoura was on September 22, 2006, when the final M-V rocket delivered Hinode (Solar B) into Earth orbit.

Vandenberg Air Force Base covers 400 km^2 (150 mi^2) of former grazing land in Santa Barbara County on California's central coast. It is located 19 km (12 mi) north of Lompoc and 240 km (150 mi) northwest of Los Angeles. Vandenberg's location, with a stretch of clear ocean for thousands of miles to the south, makes it ideal for the launch of satellites into polar orbits from north to south.

Vandenberg's military service dates back to 1941, when it was an Army training facility known as Camp Cooke. Since 1956, it has been responsible for missile and space launches on the US West Coast. In 1957, the facility was transferred to the US Air Force and acquired its present name. In the decades since, it has served as a staging ground for ballistic missile tests as well as the launch of space-bound rockets. Today, the base is operated by the Air Force Space Command's 30th Space Wing.

Vandenberg is the only military installation in the continental United States from which unmanned government and commercial satellites are launched into polar orbit. The base sends satellites to polar orbits by launching them due south and also test-fires America's ICBMs westward toward the Kwajalein Atoll in the Marshall Islands.

The base has launch pads for Atlas, Delta, Taurus, and other rockets. The air-launched Pagasus XL also flies from Vandenberg. The northern area is generally employed for missile development and operational test launches; most are deactivated, but Minuteman and Peacekeeper ICBM silos and pads remain in use. Apart from the SLC-2W Delta pad, the orbital pads are clustered in the Point Arguello area in the south. The commercial California Spaceport, created in 1995, is also located at the southern end of the base. The site is leased from US Air Force for 25 years.

Until 1994, Scout rockets were launched from Vandenberg's complex 5. Titan vehicles were supported by SLC-4 until October 19, 2005. Vandenberg was to have been a second launch site for the Space Shuttle, but the facilities were never completed. The former shuttle SLC-6 pad is now used for Delta IV launches. Vandenberg supports a population of over 18,000 (military, family members, contractors, and civilian employees).

Wallops Flight Facility, Virginia USA

37.8°N, 75.5°W

Wallops Flight Facility, located on the eastern shore of Virginia about 240 km (150 mi) southeast of Washington, is one of the oldest launch sites in the world. The site was founded in 1945 as a National Advisory Committee for Aeronautics (NACA) facility, and the first suborbital firing made from Wallops was a Tiamat on July 4, 1945. It became America's third orbital launch site in 1961 with the launch of Explorer 9 on a solid-fuel Scout rocket. Nineteen orbital launches were conducted until the end of 1985, when NASA and the USAF largely suspended orbital operations. The Scout was retired in 1994, and an attempt to launch the commercial Conestoga rocket failed in 1995.

The Virginia Commercial Space Flight Authority, a state agency created in 1995, built a Minotaur launch pad in 1998 on land leased from NASA on the south end of Wallops Island. Now known as the Mid-Atlantic Regional Spaceport, this commercial facility has two launch pads, known as Launch Complex Zero, capable of launching small and medium vehicles to LEO. Two Minotaur launches took place in December 2006 and April 2007.

Wallops is comprised of three separate areas: the Mainland, the Main Base, and Wallops Island. Over the years, the range has grown to six launch pads with associated assembly, control, and tracking facilities. Today, it is still the main center for NASA's sounding rocket program, using Black Brant, Taurus-Tomahawk, Taurus-Orion, and Terrier-Malemute rockets. Wallops conducted 11 sounding-rocket launches in 2003, but has completed only a few in each of the years since then.

Wallops offers a wide array of launch vehicle trajectory options. In general, it can accommodate launch azimuths between 90° and 160°. Most orbital launches occur between 38° and approximately 60°. Other trajectories, including polar and Sun-synchronous orbits, can be achieved by in-flight azimuth maneuvers.

From 1945 to 1957, Wallops was known as the Pilotless Aircraft Research Station. From the birth of NASA in 1958 until 1974, it was known as Wallops Station. From 1975–1981, the site was called Wallops Flight Center. Since 1982, it has been called Wallops Flight Facility, and today it is part of NASA's Goddard Space Flight Center. Wallops employs about 1000 civil service and contractor staff.

Xichang CHINA

28.25°N, 102.0°E

Xichang is the most modern of China's three main launch centers. Located 65 km (40 mi) north of Xichang City, Sichuan Province, in southwestern China, it is further south than the others (although a new site is under development on Hainan Island) and so offers better access to geostationary orbit. Its lies about 1839 m (6030 ft) above sea level and has a subtropical climate, which means temperatures are generally very mild in all seasons. The dry season (October–May), is the most suitable time for launch campaigns, but the site enjoys 320 days of sunshine annually.

Construction began in 1978, with the first launch in January 1984. The center has two launch pads for the launch of geostationary communications satellites and meteorological satellites by Long March rockets. One pad is designed for Long March 3 rockets. The second complex, 300 m (985 ft) away, is used by Long March 2E/3A/3B vehicles fitted with add-on boosters. The first launch of a Long March 2E took place on July 16, 1990, delivering the Pakistani Badr-A scientific satellite and a Chinese satellite into orbit.

The area is fairly densely populated, and launch accidents have resulted in civilian casualties. When the first Long March 3B rocket crashed on a hillside 1.6 km (1 mi) from the launch pad in 1996, six people were killed and 57 injured. The launch failure of a Long March 2E in 1995 killed six and injured 23 in a village 8 km (5 mi) downrange.

The site's first commercial communications mission was in April 1990, when the Hong Kong–owned AsiaSat 1 was launched on a Long March 3. Recent launches include the first of the Double Star satellites and the Beidou navigation satellites.

The site lies 50 km from Xichang Airport and has links to the national Chengdu–Kunming railway and the Sichuan–Yunnan highway. Launcher stages are delivered by rail to a transit hall and assembly room capable of processing three complete vehicles. The launchers are assembled, checked out, and then transferred separately by road for stacking in stages on the pad. The command and control center lies 7 km (4.3 mi) southeast of the pads.

Civil Communi-
cations &
Applications
Satellites

Americom (AMC) USA

Communications

The largest supplier of satellite services in the Americas, SES AMERICOM was established in 1973 with its first satellite circuit for the US DOD. The company currently operates a fleet of 16 spacecraft, predominantly providing service throughout the Americas.

AMC-18 provides advanced C-band digital transmission services, including high-definition channels, for cable programming and broadcasting in the continental US, Mexico, and the Carribean. The three-axis-stabilized satellite is based on the Lockheed A 2100 platform. Its dry mass is 918 kg (2020 lb) and its span in orbit 14.7 m (48 ft). AMC-18 has onboard power of 1467 W, with a projected lifetime of at least 15 years.

SPECIFICATION (AMC-18):

MANUFACTURER: Lockheed Martin Commercial Space Systems
LAUNCH DATE: December 8, 2006
ORBIT: 105°W (GEO)
LAUNCH SITE: Kourou, French Guiana
LAUNCHER: Ariane 5 ECA
LAUNCH MASS: 2081 kg (4578 lb)
BODY DIMENSIONS: 3.8 x 1.9 x 1.9 m (12.5 x 6.2 x 6.2 ft)
PAYLOAD: 24 C-band transponders

AMOS ISRAEL

Communications

SPECIFICATION
(AMOS-2):

MANUFACTURER: Israel Aircraft
Industries
LAUNCH DATE: December 28,
2003
ORBIT: 4°W (GEO)
LAUNCH SITE: Baikonur,
Kazakhstan
LAUNCHER: Soyuz-Fregat
LAUNCH MASS: 1374 kg (3023 lb)
BODY DIMENSIONS: 2.7 x 2.1 x
2.4 m (8.9 x 6.9 x 7.9 ft)
PAYLOAD: 12 Ku-band
transponders

The AMOS class is a family of lightweight geosynchronous communication satellites, developed, launched, and controlled by IAI/MBT. The satellites, colocated at 4°W, provide high-quality broadcasting and communications services to Europe, the Middle East, and the US East Coast.

AMOS-Spacecom operates AMOS-1 and AMOS-2 in orbit, and intends to launch AMOS-3 in the near future. AMOS-1, launched in 1996, was followed by AMOS-2 in 2003. AMOS-2 expands the available bandwidth and coverage, offering a hot spot for European and Middle East customers. It also adds a third beam covering the Atlantic bridge from the US East Coast to its other service areas, creating a vital link for VSAT operators, government agencies, and other service providers.

AMOS-2 has 22 active 36 MHz segments, while AMOS 1 has 14. AMOS 2 serves more transponders, and each of these provides 76 W of power, compared with 33 W for AMOS 1. AMOS-2 was the Soyuz rocket's first commercial launch to GEO; it was originally scheduled for launch on board Ariane 5. AMOS-3 is scheduled to be launched by the end of 2007, eventually replacing AMOS-1.

Anik (Telestar Canada) CANADA

Communications

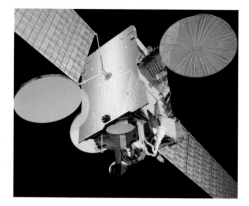

SPECIFICATION
(Anik F2):

MANUFACTURER: Boeing Satellite Systems
LAUNCH DATE: July 17, 2004
ORBIT: 111.1°W (GEO)
LAUNCH SITE: Kourou, French Guiana
LAUNCHER: Ariane 5G
LAUNCH MASS: 5950 kg (13,090 lb)
BODY DIMENSIONS: 7.3 x 3.8 x 3.4 m (23.9 x 14.4 x 11.2 ft)
PAYLOAD: 38 Ka-band transponders; 32 Ku-band transponders; 24 C-band transponders

Canada's Telesat is one of the world's leading commercial satellite operators. Created in 1969, the company made history in 1972 with the launch of Anik A1—the world's first commercial domestic communications satellite placed in geostationary orbit. In the Inuit dialect, "Anik" means "little brother."

Today, the company has eight satellites in GEO slots over North America. The most recent of these is Anik F3, which was launched on April 9, 2007. Its predecessor, Anik F2, was the world's largest commercial communications satellite and the first to fully commercialize the Ka-frequency band for delivering two-way broadband services.

Anik F2 is used to deliver high-speed satellite Internet services to consumers and businesses in North America, as well as direct-to-home satellite television and interactive services for the Canadian government, including interactive healthcare and learning, and e-government services. Based on the Boeing 702 bus, it carries 14 reflectors: two dual-gridded reflectors (one each in C-band and Ku-band), four 1.4 m (4.6 ft) transmit reflectors in Ka-band, four 0.9 m (3 ft) receive reflectors also in Ka-band, and two 0.5 m (1.6 ft) track reflectors in Ku-band. It is expected to operate for at least 15 years.

APTSTAR (APT) HONG KONG/CHINA

Communications

SPECIFICATION
(APSTAR-VI):

MANUFACTURER: Alcatel Space
LAUNCH DATE: April 12, 2005
ORBIT: 134°E (GEO)
LAUNCH SITE: Xichang, China
LAUNCHER: Long March 3B
LAUNCH MASS: 4680 kg
(10,310 lb)
BODY DIMENSIONS: 4 m (13.1 ft)
long
PAYLOAD: 38 C-band
transponders; 12 Ku-band
transponders

APT Satellite Holdings of Hong Kong successfully launched its first Comsat, APSTAR-1, in 1994. Today, APT operates five satellites in GEO, providing coverage over Asia, Oceania, and parts of the Pacific Ocean as far as Hawaii. With their high-power transmission and broad footprints, the satellites' services reach over 70% of the world's population.

APSTAR-VI is based on the SPACEBUS-4100 C1 platform developed by Alcatel Space. The C-band transponders cover Asia, Australia, New Zealand, the Pacific Islands, and Hawaii. The Ku-band transponders mainly focus on the Greater China region. Designed with powerful transponders, single polarization, and other unique features, APSTAR-VI is used for direct-to-home and other broadcasting and telecommunication services. It is the first commercial satellite in China to be equipped with an antijamming feature. It has an estimated operational lifetime of over 15 years.

169

Arabsat SAUDI ARABIA

Communications

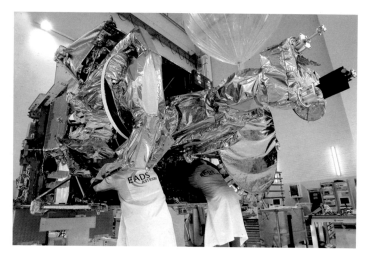

Arabsat was founded in 1976 by the 21 member states of the Arab League. It is the leading satellite services provider in the Arab world, reaching over 100 countries across the Middle East, Africa, and Europe. It operates a fleet of four satellites at the 26°E and 30.5°E slots in GEO, with two more satellites planned. Arabsat is the only satellite operator in the region offering the full spectrum of broadcast, telecommunications, and broadband services.

Arabsat's fourth-generation satellites, Arabsat-4A and Arabsat-4B, are based on the Eurostar-2000+ bus. Arabsat-4A, also known as Badr-1, was lost during an upper-stage Proton/Breeze M malfunction in 2006. Its replacement, called Arabsat-4AR/Badr-6, is scheduled for launch in 2008. Arabsat-4B/Badr-4 was successfully launched and carries direct-to-home television services, together with voice and data telecommunications services. The 25 m (82 ft) solar-array span produces 5 kW of power. It is colocated at 26°E with Arabsat-A/Badr-3.

SPECIFICATION (Badr-4/Arabsat-4B):

MANUFACTURER: EADS Astrium
LAUNCH DATE: November 8, 2006
ORBIT: 26°E (GEO)
LAUNCH SITE: Baikonur, Kazakhstan
LAUNCHER: Proton/Breeze M
LAUNCH MASS: 3280 kg (7216 lb)
BODY DIMENSIONS: 1.8 x 2.3 m (5.9 x 7.5 ft)
PAYLOAD: 32 Ku-band transponders

Artemis (Advanced Relay and Technology Satellite) EUROPE

Experimental optical communications

Artemis was built to demonstrate new techniques for data relay and mobile communications. It was also the first ESA spacecraft to carry operational electric propulsion. Due to a malfunction of the Ariane 5 upper stage, it was left in a low transfer orbit of 590 x 17,487 km (367 x 10,866 mi). To recover the mission, its liquid apogee engine was fired eight times. From a circular 31,000 km (19,263 mi) orbit, its ion thrusters were used to reach operational altitude in GEO on January 31, 2002. Enough fuel/xenon remained for up to 10 years of operations.

On November 21, 2001, it completed the first laser data transmission between European satellites, through an optical data link with SPOT 4, which was in LEO at 832 km (517 mi). On December 9, 2005, the first ever bidirectional optical interorbit communication was made using a laser beam between Japan's Kirari (OICETS) and Artemis. On December 18, 2006, Artemis relayed the first optical laser links from an aircraft over a distance of 40,000 km (24,855 mi) during two flights.

SPECIFICATION:

MANUFACTURER: Alenia Spazio
LAUNCH DATE: July 12, 2001
ORBIT: 21.5°E (GEO)
LAUNCH SITE: Kourou, French Guiana
LAUNCHER: Ariane 5
LAUNCH MASS: 3105 kg (6831 lb)
DIMENSIONS: In orbit, 4.8 x 8 x 25 m (15.7 x 26.2 x 82 ft)
PAYLOAD: S/Ka-band Data Relay (SKDR); Semiconductor Laser Intersatellite Link Experiment (SILEX); SILEX and SKDR feeder link; L-band Land Mobile (LLM); Navigation payload (NAV)

AsiaSat HONG KONG/CHINA

Communications

Asia Satellite Telecommunications Company (AsiaSat) was formed in 1988 as Asia's first privately owned regional satellite operator. AsiaSat 1 was launched on April 7, 1990. There are currently three operational satellites—AsiaSat 2, AsiaSat 3S, and AsiaSat 4—that are designed to cover the Asia-Pacific region. The latest of these, AsiaSat 4, is a Boeing 601HP (High Power) satellite that offers excellent "look angles" over Asia and Australasia.

AsiaSat 4 generates up to 9600 W using two sun-tracking, four-panel solar wings covered with triple-junction gallium arsenide solar cells. It has 28 linearized C-band and 20 Ku-band transponders, with 16 operating in the FSS (Fixed Satellite Service) and 4 operating in BSS (Broadcasting Satellite Service) frequency band. It offers extensive C-band coverage across Asia, with targeted Ku-band beams for Australasia and East Asia, and a BSS payload for the provision of direct-to-home services to Hong Kong. In April 2006, AsiaSat awarded a contract to build AsiaSat 5 as a replacement for AsiaSat 2. Launch is scheduled for 2009.

SPECIFICATION
(Asiasat 4):

MANUFACTURER: Boeing Satellite Systems
LAUNCH DATE: April 11, 2003
ORBIT: 122.2°E (GEO)
LAUNCH SITE: Cape Canaveral, Florida
LAUNCHER: Atlas IIIB
LAUNCH MASS: 4137 kg (9121 lb)
BODY DIMENSIONS: 5.8 x 3.4 x 3.5 m (19 x 11.1 x 11.4 ft)
PAYLOAD: 28 C-band transponders; 20 Ku-band transponders

SES ASTRA is the leading direct-to-home satellite system in Europe, delivering satellite TV and cable services to more than 109 million households. The ASTRA satellite fleet currently comprises 12 satellites, transmitting 1864 analog and digital television and radio channels. SES ASTRA also provides satellite-based multimedia, Internet, and telecommunication services to enterprises, governments, and their agencies.

ASTRA 1L is located at one of Astra's three prime orbital positions, replacing Astra 2C, which was moved to 28.2°E slot. It is used to provide distribution of direct-to-home broadcast services across Europe as well as a Ka-band payload for interactive applications. ASTRA 1L is a three-axis-stabilized A2100 AXS relay platform built by Lockheed Martin. The 27 m (88.6 ft) span solar arrays produce about 13 kW of power, and are expected to provide 15 years of design life.

SPECIFICATION (ASTRA 1L):

MANUFACTURER: Lockheed Martin Commercial Space Systems
LAUNCH DATE: May 4, 2007
ORBIT: 19.2°E (GEO)
LAUNCH SITE: Kourou, French Guiana
LAUNCHER: Ariane 5-ECA
LAUNCH MASS: 4497 kg (9893 lb)
BODY DIMENSIONS: 7.7 x 2.6 x 3.6 m (25.3 x 8.6 x 11.9 ft)
PAYLOAD: 29 Ku-band transponders; 2 Ka-band transponders

Brasilsat BRAZIL

Communications

SPECIFICATION
(Brasilsat B4):

MANUFACTURER: Hughes Space
and Communications
LAUNCH DATE: August 21, 2000
ORBIT: 92°W (GEO)
LAUNCH SITE: Kourou, French
Guiana
LAUNCHER: Ariane 44 LP
LAUNCH MASS: 1757 kg (3865 lb)
BODY DIMENSIONS: 3.4 x 3.7 m
(11.3 x 12 ft)
PAYLOAD: 28 C-band
transponders

Embratel, a Brazilian communications company, pioneered satellite communications in the country with the launch of Brasilsat A1 in 1985, creating the first domestic satellite system in Latin America. Brasilsat A2 was launched in 1986, followed by the first second-generation satellite, Brasilsat B1, in 1994. Since then, three additional B-series satellites have been launched.

Brasilsat B4, which replaced Brasilsat A2, carries voice and data traffic. It is a Hughes HS 376W (Wide-body) spin-stabilized satellite with two cylindrical, telescoping solar panels covered with large-area silicon solar cells that provide up to 1800 kW of power. The bottom panel slides over the upper portion for launch and the antenna folds down. Deployed in orbit, the spacecraft extends to a height of 8.3 m (27.3 ft) and weighed 1052 kg (2320 lb) at the start of orbital operations. The satellite has a design life of 12 years. Two follow-on spacecraft, known as Star One C1 and C2, are expected to replace Brasilsat B1 and B2.

B-SAT JAPAN

Communications

Japan's Tokyo-based Broadcasting Satellite System Corporation (B-SAT) was established in 1994. The company currently owns and manages five satellites: BSAT-1a and-1b for analog services; BSAT-2a and -2c for digital services; and BS-3N, built by Lockheed Martin, as a spare. BSAT-3a is scheduled for launch in late 2007.

BSAT-2c is the third a series of Orbital Sciences satellites built for B-SAT since 1999. It replaced BSAT-2b, which was lost during a launch failure in July 2001. Based on the company's STAR platform, the three-axis-stabilized spacecraft has two solar arrays spanning 16 m (52.5 ft) and generating 2600 W. It provides direct digital television broadcast links throughout Japan, with the capacity to reach 13 million homes. The spacecraft's lifetime is 10 years.

SPECIFICATION (BSat-2c):

MANUFACTURER: Orbital Sciences Corporation
LAUNCH DATE: June 11, 2003
ORBIT: 110ºE (GEO)
LAUNCH SITE: Kourou, French Guiana
LAUNCHER: Ariane 5G
LAUNCH MASS: 1290 kg (2838 lb)
BODY DIMENSIONS: 3.8 x 2.5 x 2 m (12.3 x 8.2 x 6.7 ft)
PAYLOAD: 8 Ku-band transponders

ChinaSat (Zhongxing) CHINA

Communications

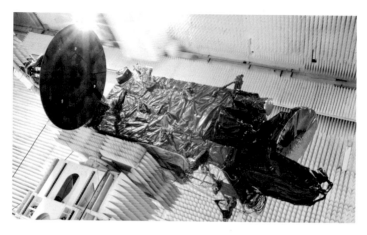

China Telecommunications Broadcast Satellite (ChinaSat) began in 1983 as a subsidiary of China's Ministry of Post and Telecommunications. It later became a subsidiary of China Telecom. In 2001, it was re-created as the China Satellite Communications Corporation. Its business includes civil fixed and mobile telecommunications and television broadcast services in China. At first, the program was purely military. Chinasat-5/Spacenet-1, launched in 1984, was the first civilian satellite.

ChinaSat has purchased and operated both Chinese-built DFH and western satellites. One of these, ChinaSat-8, was ordered in the late 1990s from Space Systems/Loral but grounded when the US refused to issue an export license. The most recent satellite is ChinaSat-6B (Zhongxing-6B), which is based on the Thales Alenia Space 4000C2 platform and produces 8.7 kW of power. It will enable China Satcom to expand its TV and fixed-communication services. The ChinaSat-9 (Zhongxing-9) satellite, also built by Thales Alenia, is the country's second direct-to-home satellite. It is scheduled for launch in late 2007.

SPECIFICATION
(Chinasat-6B):

MANUFACTURER: Thales Alenia Space
LAUNCH DATE: July 5, 2007
ORBIT: 115.5°E (GEO)
LAUNCH SITE: Xichang, China
LAUNCHER: Long March 3B
LAUNCH MASS: 4600 kg (10,120 lb)
PAYLOAD: 38 C-band transponders

DIRECTV USA

Satellite TV broadcasting

SPECIFICATION (DIRECTV 10):

MANUFACTURER: Boeing Space and Intelligence Systems
LAUNCH DATE: July 7, 2007
ORBIT: 102.8°W (GEO)
LAUNCH SITE: Baikonur, Kazakhstan
LAUNCHER: Proton/Breeze M Enhanced
LAUNCH MASS: 5893 kg (12,965 lb)
BODY DIMENSIONS: 8 x 3.7 x 3.3 m (26.2 x 12.1 x 10.8 ft)
PAYLOAD: 44 Ka-band transponders

DIRECTV of El Segundo, California, delivers satellite-based TV services to over than 15 million US homes and businesses. The company currently has 10 satellites in its constellation. The latest of these is DIRECTV 10, the heaviest and most powerful DIRECTV spacecraft so far launched.

Built on the Boeing 702 bus, it has a 48 m (175 ft) solar array, made of triple-junction gallium arsenide solar cells, which provide 18 kW of power. It carries two 2.8 m (9.2 ft) Ka-band reflectors and nine other Ka-band reflectors. The payload integrates 32 active and 12 spare National Service Ka-band Traveling Wave Tube Amplifiers (TWTAs) in addition to 55 active and 15 spare spot-beam TWTAs. It is designed to provide local and national High-Definition Television (HDTV) throughout the US from a prime orbital slot over the east Pacific. Its projected life span is 15 years. Two similar satellites have been ordered.

ETS-VIII (Engineering Test Satellite, KIKU No. 8) JAPAN

Experimental satellite communications

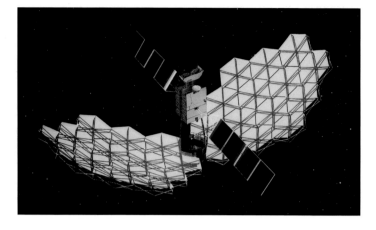

Japan's space agency has been launching a series of Engineering Test Satellites (ETS series) since the launch of ETS-I (KIKU No. 1) in 1975, with the aim of developing and testing advanced satellite technologies. The most recent is Engineering Test Satellite VIII (KIKU No. 8), one of the largest geostationary satellites ever launched. ETS-VIII will conduct orbital experiments on the Large-scale Deployable Reflector (for S-band). The satellite carries two of the largest deployable antennas in the world and two solar panels that generate 7.5 kW. Its overall length is 40 m (131 ft). The area of each deployable mesh reflector is 19 x 17 m (62.3 x 55.8 ft). The dishes are made from gold-plated molybdenum. Soon after deployment, a problem was traced to a harness that connects the amplifiers with their power supply; however, the start of normal operations was announced in May 2007.

ETS-VIII is also equipped with high-accuracy atomic clock and time transfer equipment. Combining the clock signal with GPS data, it can conduct satellite-positioning experiments. Design lifetime is 10 years.

SPECIFICATION:

MANUFACTURER: Mitsubishi Electric, NEC Toshiba Space Systems
LAUNCH DATE: December 18, 2006
ORBIT: 146ºE (GEO)
LAUNCH SITE: Tanegashima, Japan
LAUNCHER: H-IIA
LAUNCH MASS: 5800 kg (12,760 lb)
BODY DIMENSIONS: 7.3 x 2.5 x 2.4 m (23.9 x 8 x 7.7 ft)
PAYLOAD: 32 low noise amplifiers; S-band reflector; high-power transponder; atomic clock

Europestar-1 (PanAmSat-12, Intelsat-12) INTERNATIONAL

Communications

EuropeStar was a joint venture between Loral Space and Communications and Alcatel Spacecom. Based in the UK, EuropeStar was also a member of the Loral Global Alliance, a worldwide satellite network including Loral Skynet, Satmex, and Skynet do Brasil. The company originally filed rights to three orbital slots (43, 45, and 47.5°E). EuropeStar-B (formerly Koreasat 1) was eventually located at 47.5°E while EuropeStar-1 was placed in the 45°E slot, where it could serve a market of 3 billion people in 76 countries.

EuropeStar-1 was built by prime contractor Alcatel Space, which provided the payload; Space Systems/Loral, which provided the SS/L 1300 platform, assembled and tested the satellite. The powerful 12 kW satellite became the first European Ku-band satellite system to directly cover much of three continents, providing telecommunications, high-speed Internet protocol multicasting, and television distribution. In July 2005, it was sold to PanAmSat and renamed PAS-12. In February 2007, it was renamed Intelsat-12.

SPECIFICATION
(EuropeStar-1):

MANUFACTURERS: Alcatel Space, Space Systems/Loral
LAUNCH DATE: October 29, 2000
ORBIT: 45°E (GEO)
LAUNCH SITE: Kourou, French Guiana
LAUNCHER: Ariane 44LP
LAUNCH MASS: 4150 kg (9130 lb)
BODY DIMENSIONS: 2.7 x 5.5 x 3.5 m (8.9 x 18 x 11.5 ft)
PAYLOAD: 30 Ku-band transponders

Eutelsat INTERNATIONAL

Communications

Eutelsat Communications is the holding company of Eutelsat S.A. With capacity on 20 satellites (plus four with leased capacity) that provide coverage over the entire European continent, as well as the Middle East, Africa, India, and significant parts of Asia and the Americas, Eutelsat is one of the world's three leading satellite operators in terms of revenues.

One of the most recent additions to the fleet is W3A, based on the Astrium Eurostar 3000 platform. Covering Europe, the Middle East, and North Africa, the satellite almost doubled Eutelsat's capacity over sub-Saharan Africa, providing broadband services such as voice over IP, Internet access, and distance-learning services. It carries 58 transponders (of which 50 can be simultaneously operated). Up to eight transponders are available for two-way communications between Europe and Africa. Communications over Africa are in Ku-band and over Europe in Ka-band, with switching managed by the satellite.

SPECIFICATION (W3A):

MANUFACTURER: EADS Astrium
LAUNCH DATE: March 16, 2004
ORBIT: 7°E (GEO)
LAUNCH SITE: Baikonur, Kazakhstan
LAUNCHER: Proton/Breeze M
LAUNCH MASS: 4200 kg (9240 lb)
BODY DIMENSIONS: 5.8 x 2.4 x 2.9 m (19 x 7.9 x 9.5 ft)
PAYLOAD: 3 Ka-band transponders; 55 Ku-band transponders

Express RUSSIA

Communications

The Express-AM series is operated on behalf of the Russian state by the Russian Satellite Communications Co. (RSCC).The first satellite was launched on October 27, 1999. In January 2007, five satellites in the Express-AM series were operating (at 40, 53, 80, 96.5, and 140°E). The most recent of these, Express-AM3, was developed for RSCC by NPO PM, in cooperation with Alcatel Space (France).

Express-AM3 has steerable antennas that enable the satellite to serve customers in Siberia, the Far East, Asia, and Australasia with high-quality digital TV and radio broadcasting, multimedia, and data transmission services. The L-band transponder on Express-AM3 is set aside for mobile presidential and governmental communications. Its service life is 12 years.

The failure of Express AM-11 in May 2006 was attributed to an impact with space debris or a micrometeorite. The Express AM-33 and AM-44 satellites are scheduled for launch in late 2007 or 2008. RSCC is also planning to use the Express-MD series; these smaller satellites, developed by Khrunichev, are said to offer more flexibility.

SPECIFICATION
(Express-AM3):

MANUFACTURER: NPO PM
LAUNCH DATE: June 24, 2005
ORBIT: 140°E (GEO)
LAUNCH SITE: Baikonur, Kazakhstan
LAUNCHER: Proton K
LAUNCH MASS: 2600 kg (5720 lb)
BODY DIMENSIONS: not known
PAYLOAD: 16 C-band transponders; 12 Ku-band transponders; 1 L-band transponder

Foton RUSSIA

Microgravity research

The Foton unmanned spacecraft is based on the Vostok and the Zenit military reconnaissance satellite. It consists of a battery module, a spherical reentry capsule, and a service module. When introduced in 1985, Foton was mainly used for materials science research, but later missions included experiments in fluid physics, biology, and radiation dosimetry.

The improved Foton-M has larger battery capacity, enhanced thermal control, and telemetry and telecommand capabilities for increased data flow. Missions typically last 12–18 days. The 12-day Foton-M3 mission carried a European payload of 400 kg (882 lb) with experiments in fluid physics, biology, crystal growth, meteoritics, radiation dosimetry, and exobiology. It included studies of bacteria, newts, geckos, fish, and snails. The BIOPAN-6 platform on its exterior exposed its contents directly to the space environment. Foton-M3 also carried YES2, which released a small Fotino capsule at the end of an 30 km (18.6 mi) tether.

SPECIFICATION (Foton-M3):

MANUFACTURER: TsSKB-Progress
LAUNCH DATE: September 14, 2007
ORBIT: 258 x 280 km (160 x 174 mi), 62.9° inclination
LAUNCH MASS: 70 kg (154 lb)
LAUNCH SITE: Baikonur, Kazakhstan
LAUNCHER: Soyuz U
BODY DIMENSIONS: Approx. 6.5 x 2.5 m (21.3 x 8.2 ft)
PAYLOAD: ESA microgravity payload, including BIOPAN-6; Second Young Engineers' Satellite (YES2)

Galaxy INTERNATIONAL

Communications

SPECIFICATION
(Galaxy 17):

MANUFACTURER: Thales Alenia
Space
LAUNCH DATE: May 4, 2007
ORBIT: 91° W (GEO)
LAUNCH SITE: Kourou, French
Guiana
LAUNCHER: Ariane 5-ECA
LAUNCH MASS: 4100 kg (9020 lb)
BODY DIMENSIONS: 3.8 x 1.8 x
2.3 m (12.3 x 5.9 x 7.5 ft)
PAYLOAD: 24 C-band
transponders; 24 Ku-band
transponders

The Galaxy series of communications satellites was originally owned and operated by Hughes Communications. The first three satellites were launched in 1983–4. The company was later sold to PanAmSat, which in turn merged with Intelsat in 2006. On February 1, 2007, when Intelsat changed the names of 16 of its satellites, "Galaxy" replaced the "Intelsat Americas" brand.

The latest of 16 operational satellites in the series is Galaxy 17, which uses a Thales Alenia Spacebus 3000 B3 platform. Eventually to be located at 91°W, it takes over the orbital slot of Galaxy 11, a Boeing 702 model launched in 1999, whose solar arrays have a defect that results in a gradual loss of power. The dedicated cable satellite will join Intelsat's fleet that serves video, corporate, and consumer broadband customers in North America, Central America, and the Caribbean. The three-axis-stabilized spacecraft has two solar arrays that span 36.9 m (121 ft) and provide 8.6 kW of power at end of life. Its service life is 15 years.

Garuda 1 (ACeS) INDONESIA

Communications

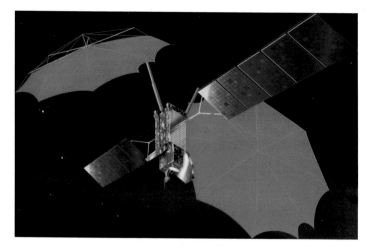

The ACeS (ASIA Cellular Satellite) Garuda 1 was the first regional, satellite-based mobile telecommunications system specifically designed for the Asian market. It is owned by a consortium including Lockheed Martin and companies from the Philippines, Indonesia, and Thailand.

Garuda 1 uses the Lockheed Martin A2100XX spacecraft bus. The 14 kW satellite was one of the most powerful commercial Comsats ever produced and the heaviest single Comsat ever launched on a Proton. However, an anomaly with one of the two L-band antennas reduces the maximum call capacity from 2 million to 1.4 million a day.

With a projected service life of 12 years, the payload features two 12 m (39.4 ft) L-band antennas, which beam signals directly to dual-mode handheld phones via 140 satellite spot beams. A second satellite is planned as a back-up at 118°E, but in 2006 ACeS agreed to combine coverage of the Garuda 1 satellite and the Inmarsat 4 series of satellites.

SPECIFICATION:

MANUFACTURER: Lockheed Martin Commercial Space Systems
LAUNCH DATE: February 12, 2000
ORBIT: 123° E (GEO)
LAUNCH SITE: Baikonur, Kazakhstan
LAUNCHER: Proton K/Block DM
LAUNCH MASS: 4500 kg (9900 lb)
BODY DIMENSIONS: 6 m (19.7 ft) max.
PAYLOAD: 88 (+22) L-band transponders

Genesis USA

Inflatable spacecraft research

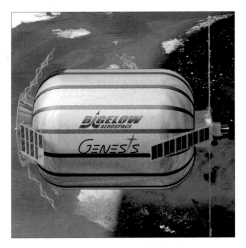

SPECIFICATION
(Genesis I):

MANUFACTURER: Bigelow
Aerospace
LAUNCH DATE: July 12, 2006
ORBIT: 550 km, 64° inclination
LAUNCH MASS: 69.8 kg (154 lb)
LAUNCH SITE: Dombarovsky/
Yasny, Russia
LAUNCHER: Dnepr
BODY DIMENSIONS: 4.4 x 2.5 m
(14.4 x 8.3 ft)
PAYLOAD: GeneBox; Life science
experiment; personal items
and photos

The two Genesis modules are one-third scale models of future inflatable spacecraft that are expected to begin flying as early as 2010. They are based on NASA's Transhab concept and developed with private funding by Bigelow Aerospace. Each module has a 15 cm (6 in) thick multilayer exterior. Once in LEO, the 4.4 m (15 ft) module deploys eight solar arrays that provide 1 kW of power and expands from its launch width of 1.9 m (6.2 ft) to 2.5 m (8.3 ft).

Genesis II, launched on June 28, 2007, is a near-twin of Genesis I, but the later version has an improved attitude-control system and a distributed multitank inflation system that replaces the single-tank design of Genesis I. Both modules carry cameras to image the exterior and record views inside the spacecraft's 11.5 m³ (406 ft³) volume. Genesis I has six internal and seven external cameras, and Genesis II has a total of 22. Data and images are sent to the mission control center in North Las Vegas, Nevada.

Genesis I contains various personal items and photos, a life science experiment that includes cockroaches, Mexican jumping beans, and a NASA experiment called GeneBox. Genesis II also carries personal items, as well as a Space Bingo game and Biobox, a three-chamber pressurized vessel with compartments for biological specimens to be observed by onboard cameras. Occupants include cockroaches, scorpions, and red harvester ants.

GIOVE (Galileo In-Orbit Validation Element)   EUROPE

Navigation

ESA and the European Community have agreed to create Galileo, the first civil satellite positioning, navigation, and timing system. Development began in 2003, including three or four satellites to test the system. The GIOVE satellites secure access to the Galileo frequencies allocated by the International Telecommunications Union, study the radiation environment, and test critical technologies (e.g., onboard atomic clocks, signal generator, and user receivers). GIOVE-A and -B were built to provide in-orbit redundancy, and their capabilities are complementary. The smaller GIOVE-A carries a rubidium atomic clock (plus backup) and transmits a signal through two separate channels at a time. The three-axis-stabilized satellite has a cube-shaped body and two solar arrays that generate 700 W. The larger Galileo Industries satellite adds a hydrogen-maser clock and will transmit a signal through three channels. GIOVE-A is expected to operate until early 2008. A third Surrey Satellite Technology test satellite, GIOVE-A2, is scheduled for launch in 2008.

SPECIFICATION (GIOVE-A):

MANUFACTURER: Surrey Satellite Technology
LAUNCH DATE: December 28, 2005
ORBIT: 23,226 x 23,285 km (14,432 x 14,469 mi), 56° inclination
LAUNCH MASS: 600 kg (1320 lb)
LAUNCH SITE: Baikonur, Kazakhstan
LAUNCHER: Soyuz/Fregat
BODY DIMENSIONS: 1.3 x 1.8 x 1.7 m (4.3 x 5.9 x 5.4 ft)
PAYLOAD: 2 rubidium atomic clocks; L-band antenna; signal generation units; 2 radiation monitors; navigation receiver

Globalstar USA

LEO communications constellation

SPECIFICATION
(FM65, 69, 71 and 72):

MANUFACTURER: Alcatel Alenia
Space (now Thales Alenia Space)
LAUNCH DATE: May 30, 2007
ORBIT: 920 km (572 ml), 52°
inclination (eight planes)
LAUNCH SITE: Baikonur,
Kazakhstan
LAUNCHER: Soyuz/Fregat
LAUNCH MASS: 450 kg (990 lb)
BODY DIMENSIONS: 0.6 x 1.5 x
1.6 m (2 x 4.9 x 5.2 ft)
PAYLOAD: C-band; 16 L-band
cells; 16 S-band cells

Globalstar provides mobile satellite voice and data services, especially to remote areas without a cellular or landline service, through a constellation of small satellites in phased MEOs. Under the original plan, there would be 56 satellites, including eight on-orbit spares. The first launches were in 1998, and the system has been operational since 1999. After 12 were lost in a Zenit failure, the system was restricted to 48 satellites. Eight ground spares are to be launched by Soyuz in 2007 to augment the existing constellation of 40 operational satellites.

The satellites are based on the SS/Loral LS-400 platform, with a trapezoidal body and two deployable solar panels. Like "bent pipes," or mirrors in the sky, the satellites pick up signals from over 80% of the Earth's surface. Several satellites pick up a call, ensuring that it is not lost. As soon as a second satellite picks up the signal, it can take over the transmission to a gateway on the ground. A second generation of satellites is scheduled for delivery from summer 2009.

Gonets RUSSIA

Mobile/store-dump communications

The civil Gonets (Messenger) system of small communications satellites in medium Earth orbit is derived from the Strela-3 military program. Once a state-run program, it is now operated by the Gonets SatCom company. The first Gonets-D demonstration missions took place in 1992 as Kosmos 2199 and 2201. After several more test flights, six Gonets-D1 satellites were launched into two orbital planes to create an operational system by December 2001. The cylindrical satel-lites are covered in solar cells with an extended boom for gravity stabilization and cone antennas.

The new generation Gonets-D1M (or Gonets-M) transmits digital, speech, text, and video data. It is able to transmit urgent reports during environmental disasters, support areas with poor infrastructure, and provide the exchange of commercial, medical, scientific, and technological information between individual computers. Its operational lifetime is five years. Eventually there may be 36 satellites in six to eight orbital planes.

SPECIFICATION
(Gonets-M):

MANUFACTURER: NPO PM
FIRST LAUNCH: December 21, 2005
ORBIT: 1449 x 1469 km (900 x 913 mi), 82.5°E
LAUNCH SITE: Plesetsk, Russia
LAUNCHER: Kosmos 3M
LAUNCH MASS: 280 kg (616 lb)
BODY DIMENSIONS: 0.8 x 1.6 m (2.6 x 5.2 ft)
PAYLOAD: 16 C-band transponders; 1 L-band transponder; 1 Ku-band transponder

Gorizont RUSSIA

Communications

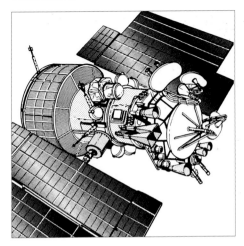

SPECIFICATION
(Gorizont 33/45L):

MANUFACTURER: NPO PM
LAUNCH DATE: June 6, 2000
ORBIT: 145°E (GEO)
LAUNCH SITE: Baikonur,
Kazakhstan
LAUNCHER: Proton K
LAUNCH MASS: 2300 kg (5060 lb)
BODY DIMENSIONS: 5.5 x 3.3 m
(17.9 x 10.8 ft)
PAYLOAD: 6 C-band transponders;
1 Ku-band transponder; 1
L-band transponder

Gorizont (Horizon) was developed during the 1970s as a successor to the Ekran Comsats. It was one of the first geostationary satellites launched by the USSR and the first Soviet civilian satellite to carry multiple transponders. It had an improved three-axis attitude control system and rotatable solar panels.

The satellites were used for data transmission, TV and radio broadcasting, telecommunications, Internet, and videoconferencing. Gorizont 1 was launched to support broadcasts of the 1980 Olympic Games from Russia. From 1988, they also supported maritime and international communications as part of the Okean system. The Gorizont system occupies 10 positions in geostationary orbit.

Three Gorizonts were deployed to an orbital location owned by Tonga and were operated for the US company Rimsat (Gorizont 17 as Tongastar 1, and Gorizont 29 and 30 as Rimsat 1 and 2). Later, the two Rimsats were transferred to PASI and AsiaSat (one each), and finally to LMI. The latest satellite, Gorizont 33, was the payload on the first successful launch of a Proton K/Breeze M rocket. Gorizont is being replaced by the Express series of satellites.

Hispasat SPAIN

Communications

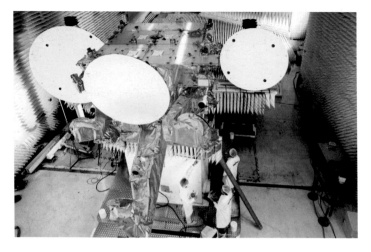

Hispasat was set up in 1989 with the objective of serving the Spanish and Portuguese language markets, including Latin America, where it is now the leading commercial satellite operator. The Hispasat 1B, 1C, and 1D satellites are located at 30°W. Other satellites in the family include Spainsat, Amazonas, and Xtar-Eur. The high-power satellites enable optimum coverage with the highest degree of flexibility in America, Europe, and North Africa.

Hispasat 1D was launched to maintain the Ku-band coverage offered through 1A and 1B. Based on the Alcatel three-axis-stabilized SPACEBUS 3000B platform, it has three antennas, 28 transponders, and connective flexibility in the Ku-band frequencies. Hispasat 1D brings the additional capacity growth of six transponders with American and transatlantic connectivity in order to meet an expected growth in demand. A beam over the Middle East gives access to Asian satellites for American and European customers. The operational lifetime of Hispasat 1D is 15 years.

SPECIFICATION (Hispasat 1D):

MANUFACTURER: Alcatel Space
LAUNCH DATE: September 18, 2002
ORBIT: 30° W (GEO)
LAUNCH SITE: Cape Canaveral, Florida
LAUNCHER: Atlas II AS
LAUNCH MASS: 3288 kg (7234 lb)
BODY DIMENSIONS: 2.8 x 2.3 x 1.8 m (9.2 x 7.5 x 5.89 ft)
PAYLOAD: 28 Ku-band transponders

Hot Bird INTERNATIONAL

Communications

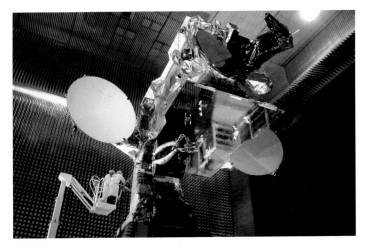

Eutelsat Communications, based in Paris, provides communications coverage over all of Europe and much of the globe. In July 2007, Eutelsat's satellites were broadcasting over 2600 TV channels and 1100 radio stations. Three Hot Bird satellites, located at 13°E, form one of the largest broadcasting systems in Europe, delivering over 1000 television channels to more than 120 million homes in Europe, North Africa, and the Middle East. In addition, the system provides multimedia services and over 600 radio stations.

A "Superbeam" enables direct-to-home reception in beam center with antennas smaller than 0.7 m (2.3 ft), and a "Widebeam" with slightly larger antennas offers reception throughout Europe, North Africa, and as far East as Moscow and Dubai. With a launch mass of 4.9 tonnes (10,803 lb), solar-array power of almost 14 kW at end of life, and a record-breaking 64 transponders, Hot Bird 8 is the largest and most powerful broadcast satellite serving Europe. The identical Hot Bird 9, based on the Eurostar E3000 platform, will follow in 2008.

SPECIFICATION
(Hot Bird 8):

MANUFACTURER: EADS Astrium
LAUNCH DATE: August 4, 2006
ORBIT: 13° E (GEO)
LAUNCH SITE: Baikonur, Kazakhstan
LAUNCHER: Proton/Breeze M
LAUNCH MASS: 4875 kg (10,748 lb)
BODY DIMENSIONS: 5.8 x 2.4 x 2.8 m (19 x 7.9 x 9.5 ft)
PAYLOAD: 64 Ku-band transponders

ICO USA

Communications

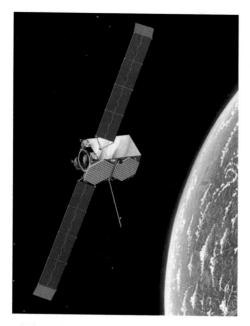

SPECIFICATION
(ICO F2):

MANUFACTURER: Hughes
(later Boeing)
LAUNCH DATE: June 19, 2001
ORBIT: 10,390 km (6456 mi),
45° inclination
LAUNCH SITE: Cape Canaveral,
Florida
LAUNCHER: Atlas IIAS
LAUNCH MASS: 2750 kg (6050 lb)
BODY DIMENSIONS: 4.7 x 2.3 x
2.3 m (15.4 x 7.5 x 7.5 ft)
PAYLOAD: 163 S-band spot
beams; C-band transponders

Virginia-based ICO Global Communications (ICO for Intermediate Circular Orbit) was established to provide global mobile personal communications services. Originally begun by Inmarsat, it became a separate company in 1995. It emerged from bankruptcy in May 2000.

ICO ordered its first 12 satellites in July 1995 and three more in September 2000. All are versions of the Hughes (now Boeing) 601 model, with some subsystems modified for operation in MEO. The first satellite was lost in a Zenit launch failure in 2000.

The new ICO satellites will be placed in two planes 90° apart, giving complete, continuous overlapping coverage of the Earth's surface. Ten will form the base constellation and the remainder will be on-orbit spares. The first satellite (F2) was launched in 2001 and provides data-gathering services for the US government. There are 10 MEO satellites in storage, most of which are in advanced stages of completion. They feature active S-band antennas capable of forming up to 490 beams for satellite–user links and C-band hardware for satellite–ground station links. The ICO G1 geostationary satellite, which will serve the USA, is due for launch in late 2007.

Inmarsat

Communications

Inmarsat is the world's leading provider of global mobile satellite communications. It owns and operates a fleet of 10 GEO satellites. Inmarsat's first wholly owned satellites, the Inmarsat-2s, were launched in the early 1990s, and the Inmarsat-3s—the first generation to use spot-beam technology—followed later in the decade.

The Inmarsat-4 series, based on the Eurostar 3000 bus, are the largest commercial telecommunications satellites ever launched and are 60 times more powerful than the Inmarsat-3. The first Inmarsat-4s were launched in 2005. A third in the series is planned. Each satellite is equipped with a global beam, 19 regional spot beams, and 228 narrow spot beams. This will eventually deliver complete mobile broadband coverage of the planet, except for the extreme polar regions. They support the Broadband Global Area Network (BGAN) for delivery of video-on-demand, videoconferencing, fax, e-mail, phone, and Local Area Network (LAN) access at speeds up to half a megabit per second. End-of-life power is 13kW.

SPECIFICATION (Inmarsat-4F2):

MANUFACTURER: EADS Astrium
LAUNCH DATE: November 8, 2005
ORBIT: 53° W (GEO)
LAUNCH SITE: Odyssey platform, Pacific Ocean
LAUNCHER: Zenit 3SL
LAUNCH MASS: 5958 kg (13,108 lb)
BODY DIMENSIONS: 7 x 2.9 x 2.3 m (23 x 9.5 x 7.5 ft)
PAYLOAD: L-band and C-band, 630 channels; navigation package

INSAT INDIA

Communications

Started in 1983, INSAT is the largest domestic communication satellite system in the Asia-Pacific region, with nine operational satellites equipped with 175 transponders. The satellites provide services in telecommunication, television broadcasting, and meteorology, including disaster warning. The program is a joint venture involving the Indian Department of Space (DOS) and the Departments of Telecommunications, Meteorology, and Radio. The DOS is responsible for the satellite and satellite operation.

The Indian-built INSAT-4B is the twelfth in the INSAT series and the second of the INSAT-4 spacecraft. The platform is derived from the INSAT-3 series. It provides television and communication services over India, beaming TV programs directly to households using small dish antennas.

SPECIFICATION (INSAT-4B):

MANUFACTURER: ISRO/ISRO Satellite Center
LAUNCH DATE: March 11, 2007
ORBIT: 93.5°E (GEO)
LAUNCH SITE: Kourou, French Guiana
LAUNCHER: Ariane 5-ECA
LAUNCH MASS: 3028 kg (6662 lb)
BODY DIMENSIONS: 3.1 x 1.8 x 2 m (10.2 x 5.8 x 6.6 ft)
PAYLOAD: 12 C-band transponders; 12 Ku-band transponders

Intelsat INTERNATIONAL

Communications

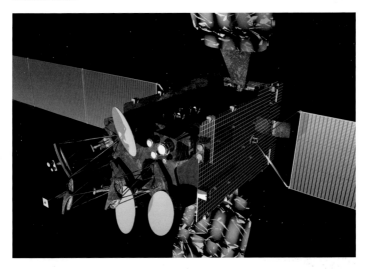

In 1965, Intelsat established the first commercial global satellite communications system. In 2006, the company merged with rival PanAmSat, creating the world's largest commercial communications satellite operator and confirming its position as the leading provider of fixed satellite services. As of July 2007, the company operated a fleet of 52 satellites under various names. The latest of these was Galaxy 17. Most of the Intelsat 7, 8, and 9 satellites are still operational.

Intelsat 10-02 is the largest and most powerful in the fleet. It has three steerable spot antennas and provides a full range of Ku- and C-band services with a high level of cross-connectivity. It provides coverage as far east as India and as far west as the Americas, with optimum footprint over Europe, Africa, and the Middle East. Its solar array span is 45 m (147.6 ft) and power at end of life will be 11 kW. The satellite is based on the Eurostar E3000 bus, and service life is at least 13 years.

SPECIFICATION (Intelsat 10-02):

MANUFACTURER: EADS Astrium
LAUNCH DATE: June 16, 2004
ORBIT: 359° E (GEO)
LAUNCH SITE: Baikonur, Kazakhstan
LAUNCHER: Proton/Breeze M
LAUNCH MASS: 5600 kg (12,320 lb)
BODY DIMENSIONS: 7.5 x 2.9 x 2.4 m (24.6 x 9.5 x 7.9 ft)
PAYLOAD: 70 C-band transponders; 36 Ku-band transponders

Iridium USA

LEO communications

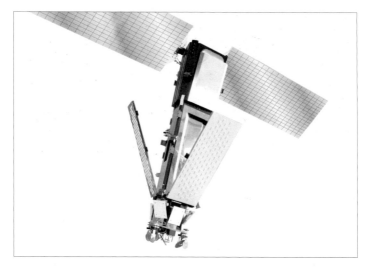

Iridium is a system of 66 active LEO communication satellites and nine spares. Since the system was originally to have 77 active satellites, it was named for the element iridium, which has atomic number 77; the original name was retained. More than 80 spacecraft were launched into LEO over an 18-month period in 1997–98, using three different boosters (Delta 7000, Proton, and Long March 2C). The first launch was on May 5, 1997. A new company, Iridium Satellite, bought the assets of bankrupt Iridium LLC in December 2000.

Lockheed Martin provided the satellite bus. The system uses L-band to provide global communications services through portable handsets. The satellites are in near-polar orbits in six planes and circle the earth once every 100 min. Each satellite is cross-linked to four others—two in the same orbital plane and two in an adjacent plane. The service has been maintained with the addition of six spares in February and May 2002, ensuring continuity until at least 2010.

SPECIFICATION (Sv97/98):

MANUFACTURERS: Motorola, Lockheed Space Systems
LAUNCH DATE: June 20, 2002
ORBIT: 780 km (484.7 mi), 75° inclination, 6 planes.
LAUNCH SITE: Plesetsk, Russia
LAUNCHER: Rockot
LAUNCH MASS: 690 kg (1520 lb)
BODY DIMENSIONS: 4 x 1.3 m (13.1 x 4.3 ft)
PAYLOAD: 1 Ka-band antenna; 3 L/S-band antennas

JCSAT JAPAN

Communications

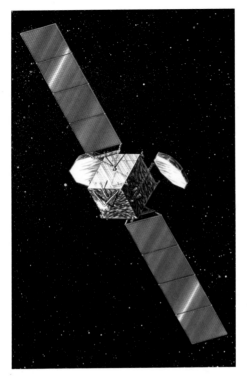

SPECIFICATION (JCSAT-10):

MANUFACTURER: Lockheed Martin Commercial Space Systems
LAUNCH DATE: August 11, 2006
ORBIT: 128° E (GEO)
LAUNCH SITE: Kourou, French Guiana
LAUNCHER: Ariane 5-ECA
LAUNCH MASS: 4048 kg (8906 lb)
BODY DIMENSIONS: 5.5 x 2.2 x 2.2 m (18 x 7.2 x 7.2 ft)
PAYLOAD: 12 C-band transponders; 30 Ku-band transponders

JSAT Corporation is a leading satellite operator in the Asia-Pacific region. The company owns and operates nine satellites in eight orbital slots covering North America, Hawaii, Asia, and Oceania. JSAT services include digital satellite TV, video, and data broadcasting.

JCSAT-10 is the second of three satellites Lockheed Martin is building for JSAT. It followed on from the identical JCSAT-9, which was launched earlier in 2006. JCSAT-11 is scheduled for launch in the third quarter of 2007.

JCSAT-10 is a high-power hybrid satellite carrying 30 active Ku-band transponders and 12 active C-band transponders that provide coverage to Japan, the Asia-Pacific region, and Hawaii. It is based on Lockheed Martin's A2100AX bus and is designed for a minimum service life of 15 years. With a span of 26.9 m (88.2 ft), its solar arrays will produce 8.7 kW at end of life.

Koreasat (Mugunghwa) SOUTH KOREA

Commercial and military communications

SPECIFICATION
(Koreasat 5):

MANUFACTURER: Alcatel Alenia Space
LAUNCH DATE: August 22, 2006
ORBIT: 113°E (GEO)
LAUNCH SITE: Odyssey platform, Pacific Ocean
LAUNCHER: Zenit 3SL
LAUNCH MASS: 4457 kg (9805 lb)
BODY DIMENSIONS: 4 x 2.2 x 2 m (13.1 x 7.2 x 6.6 ft)
PAYLOAD: 24 Ku-band transponders; 8 SHF-band transponders; 4 Ka-band transponders

Four satellites have been launched in the Koreasat series. Koreasat 1 was retired in 2005, while its two successors continue to operate (as of January 2007). The name Koreasat 4 was avoided because the number four represents death in some Asian cultures, so the fourth (and latest) satellite is actually Koreasat 5.

Koreasat 5, based on the Alcatel Alenia Spacebus 4000 C1 platform, is South Korea's first dual-use commercial and military communications satellite. Its payload of 36 transponders is divided between the co-owners, South Korea's Agency for Defense Development and the KT Corporation. Eight channels in the SHF band and four Ka-band transponders offer secure communications for the country's armed forces. Some of the technology and systems were developed for the French Defense Ministry's Syracuse 3 program.

KT Corporation is responsible for 24 Ku-band transponders that serve commercial customers in East Asia. Half of these are attached to regional beams that provide broadband multimedia and digital television services. The remainder replace domestic capacity provided by the aging Koreasat 2. In Korea, the satellite family is known as Mugunghwa (Rose of Sharon).

Measat MALAYSIA

Communications

Measat Satellite Systems uses three communications satellites to provide services to more than 70% of the world's population. The Measat fleet comprises Measat-1 (launched in 1996), Measat-2, and Measat-3. Measat-1R, under construction by Orbital Sciences Corporation, will be added in 2008.

The most recent satellite, Measat-3, is based on a Boeing 601HP satellite bus. Its two solar arrays, each with four panels of triple-junction gallium arsenide solar cells, span 26.2 m (86.0 ft) and produce 10.8 kW. It provides C- and Ku-band service to Asia, Australia, the Middle East, Eastern Europe, and Africa. Three C-band beams provide television broadcasts and telecommunications to 110 nations from Japan to Africa. Three Ku-band beams provide direct-broadcast television services to South Asia and Southeast Asia, including Indonesia and Malaysia.

SPECIFICATION
(Measat-3):

MANUFACTURER: Boeing
LAUNCH DATE: December 12, 2006
ORBIT: 91.5° E (GEO)
LAUNCH SITE: Baikonur, Kazakhstan
LAUNCHER: Proton M
LAUNCH MASS: 4765 kg (10,483 lb)
BODY DIMENSIONS: 3.8 x 7.4 m (12.5 x 24.4 ft)
PAYLOAD: 24 C-band transponders; 24 Ku-band transponders

Multi-Functional Transport Satellite (MTSAT, Himawari) JAPAN

Air traffic control and navigation/meterology

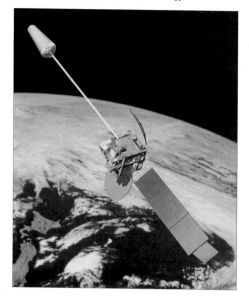

SPECIFICATION (MTSAT-2):

MANUFACTURER: Mitsubishi Electric
LAUNCH DATE: February 18, 2006
ORBIT: 145°E (GEO)
LAUNCH SITE: Tanegashima, Japan
LAUNCHER: H-IIA F9 (2024)
LAUNCH MASS: 2900 kg (6380 lb)
BODY DIMENSIONS: 4 x 2.6 x 2.6 m (13.1 x 8.5 x 8.5 ft)
PAYLOAD: Visible/infrared imaging system; S-band receive and transmit antennas; UHF antenna; Ku-band, Ka-band, and L-band communication systems

The MTSAT satellites carry payloads for weather observation and aviation control. They are owned and operated by the Japanese Civil Aviation Bureau and the Japan Meteorological Agency. Imagery is available in one visible and four infrared wavelength bands, including the water-vapor channel. The visible-light camera has a resolution of 1 km (0.6 mi); the infrared cameras have 4 km (2.5 mi) resolution.

MTSAT-1 was lost during a launch failure of the H-II rocket on November 15, 1999. MTSAT-1R (Himawari 6), also built by Space Systems/Loral, was launched by H-IIA on February 26, 2005, and is operational at 140°E. MTSAT-2 (Himawari 7) is an orbital spare, and will succeed MTSAT-1R around 2010.

The three-axis-stabilized spacecraft has a 2.7 kW gallium arsenide solar array that rotates to track the Sun. This allows the imager's north-facing passive radiation cooler to face cold space. A 3.3 m (10.8 ft) solar sail, located on a 15.1 m (49.5 ft) boom, counteracts the torque produced by sunlight pressure on the solar array. In addition to the meteorology payload, the satellite carries antennas for air traffic voice/data links and navigation.

Nahuel ARGENTINA

Communications

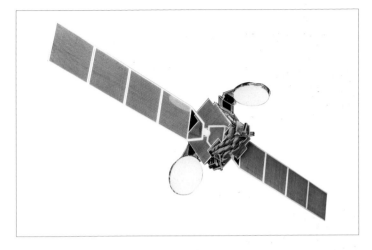

Nahuelsat S.A. is the private operator of the Argentine Nahuel Satellite System. It was set up in 1993 to operate a telecommunications system for Argentina, Chile, and Uruguay through Paracom Satelites S.A., the owner of two satellites, Nahuel C1 and C2. These were decommissioned and replaced by the currently operating Nahuel 1.

Nahuel 1 has three different beams specifically designed for each region—one for Argentina, Chile, Paraguay, and Uruguay; one exclusively for Brazil; and one for all of Latin America, the Caribbean, and the southern United States. Its functions include image and voice transmission, as well as transmission of all types of data, including Internet. Nahuel 1 is based on the Aerospatiale Spacebus 2000 platform. The solar arrays span 22.3 m (73.1 ft) and produce 3.2 kW of power. A second Nahuel satellite was canceled and a new company, ArSat, has been formed to operate and replace the current spacecraft. It was reported in July 2007 that Nahuel 1 was experiencing technical problems with its thrusters and was drifting off station.

SPECIFICATION
(Nahuel 1):

MANUFACTURER: Aerospatiale
LAUNCH DATE: January 30, 1997
ORBIT: 71.8°W (GEO)
LAUNCH SITE: Kourou, French Guiana
LAUNCHER: Ariane 44LP
LAUNCH MASS: 1780 kg (3916 lb)
BODY DIMENSIONS: 1.6 x 1.5 x 2.2 m (5.4 x 4.8 x 7.2 ft)
PAYLOAD: 18 Ku-band transponders

New Skies NETHERLANDS

Communications

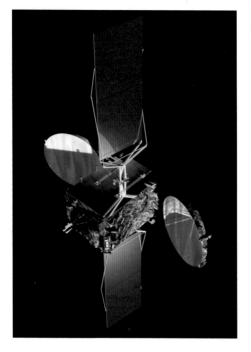

SPECIFICATION (NSS-10):

MANUFACTURER: Alcatel Space
LAUNCH DATE: February 3, 2005
ORBIT: 37.5°W (GEO)
LAUNCH SITE: Baikonur, Kazakhstan
LAUNCHER: Proton/Breeze M
LAUNCH MASS: 4979 kg (10,954 lb)
BODY DIMENSIONS: 5.1 x 2.2 x 2 m (11.2 x 7.2 x 6.6 ft)
PAYLOAD: 72 C-band transponders

SES NEW SKIES is an SES company based in the Netherlands, offering video, Internet, data, and voice communications services around the world. It operates a fleet of seven commercial geostationary communications satellites, with NSS-9 due to be added in 2009. SES NEW SKIES is complemented by SES ASTRA in Europe and SES AMERICOM in North America.

NSS-10, previously known as AMC-12/Astra 4A, was transferred to New Skies in 2007. It serves the Caribbean, South America, Europe, and Africa. It was the first satellite to be based on the Alcatel Spacebus 4000 platform. NSS-10 offers individual transponder switching capability between the Europe/Africa beam, the South America beam, and the North America beam and features "simultaneous" downlink functionality. It can also provide links to other regional satellite systems. Its three high-powered C-band beams provide high data throughput to smaller antennas with great reliability, even in areas of high rainfall. Anticipated service life is 16 years.

Nilesat EGYPT

Communications

SPECIFICATION
(Nilesat 102):

MANUFACTURER: EADS Astrium
LAUNCH DATE: August 17, 2000
ORBIT: 7° W (GEO)
LAUNCH SITE: Kourou, French
Guiana
LAUNCHER: Ariane 44LP
LAUNCH MASS: 1827 kg (4019 lb)
BODY DIMENSIONS: 2.4 x 1.7 x
2.2 m (7.9 x 5.6 x 7.2 ft)
PAYLOAD: 12 Ku-band
transponders

Nilesat is an Egyptian company established in July 1996 to operate direct-to-home broadcasting satellites and high-speed data-transmission services. It has launched two satellites into the orbital position 7°W (Nilesat 101 and Nilesat 102). Nilesat 101 was launched by an Ariane 4 in April 1998. Additional capacity is provided by Nilesat 103 (Atlantic Bird 4) which was shifted to the same orbital location.

Nilesat 102, based on the Eurostar 2000 bus, is basically identical to Nilesat 101, although it has an additional receiving horn antenna for uplinking from Europe and additional channels to those provided by Nilesat 101. Each satellite can transmit more than 100 channels of digital television to antennas only 60 cm (2 ft) in diameter. Coverage extends from Morocco to the Arabian Gulf, with high power concentrated over the most densely populated areas. The communications payload, supplied by Alcatel Space, comprises 12 Ku-band transponders and a 2.3 m (7.5 ft) diameter transmit antenna. It is expected to continue operating until 2013.

Nimiq CANADA

Communications

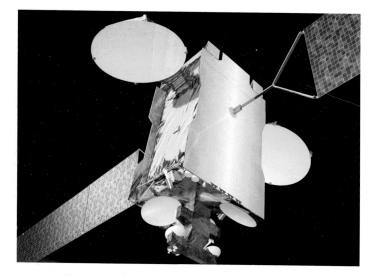

Created in 1969, Telesat Canada was one of the early pioneers in satellite communications. Today, Telesat operates a fleet of Amik and Nimiq satellites for the provision of broadcast distribution and telecommunications services.

Nimiq 1 and 2 are high-power Ku/Ka-band satellites fitted with 32 active Ku-band transponders with 120 W power amplifiers. They provide direct-to-home satellite television across Canada. Nimiq 2 also has a Ka-band payload to provide broadband services. The satellites are based on the Lockheed Martin A2100AX bus and have a minimum service life of 12 years. The name—chosen from 36,000 submissions in a national contest in 1998—is an Inuit word for any object or force that unites things or binds them together. On February 20, 2003, Nimiq 2 experienced a malfunction that reduced its available power. It is currently generating enough power to operate 26 of the 32 transponders.

SPECIFICATION (Nimiq 2):

MANUFACTURER: Lockheed Martin Commercial Space Systems
LAUNCH DATE: December 29, 2002
ORBIT: 82° W (GEO)
LAUNCH SITE: Baikonur, Kazakhstan
LAUNCHER: Proton/Breeze M
LAUNCH MASS: 3600 kg (7920 lb)
BODY DIMENSIONS: 5.8 x 2.4 x 2.4 m (19 x 7.9 x 7.9 ft)
PAYLOAD: 2 Ka-band transponders; 32 Ku-band transponders

Orbcomm USA

LEO communications constellation

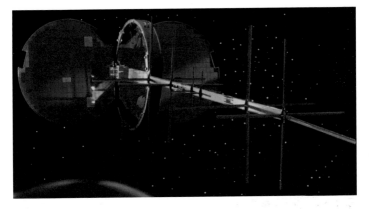

The Orbcomm system was a joint commercial venture between Orbital Communications Corporation and Teleglobe of Canada. It was the first service to provide a global, affordable two-way wireless-messaging and positioning service via handheld personal communicators and tracking units. Its purpose was to send and receive short messages anywhere on Earth at very low cost. Potential applications included real-time person-to-person or machine-to-machine communi-cations, remote industrial asset and environmental monitoring, stolen vehicle recovery and two-way e-mail communications for laptop and palmtop computers.

A small test satellite, Orbcomm-X, was launched on July 16, 1991. It had a different design and was not included in the operational system. The full Orbcomm constellation comprised 36 small satellites, including six spares, in six orbital planes. The first two satellites experienced communications problems but were recovered. These provide worldwide coverage. These satellites are small in size, since they do not require a propulsion system and have low power requirements.

SPECIFICATION:

MANUFACTURER: Orbital Sciences Corporation
FIRST LAUNCH: April 3, 1995
ORBITS: 28 satellites—800 km (497 mi), 45° inclination, 4 planes; 2 satellites—710 km and 820 km (441 and 510 mi), 70° and 108° inclination
LAUNCH SITE: Wallops Island (air launch)
LAUNCHERS: Pegasus and Pegasus XL
LAUNCH MASS: 42 kg (92 lb)
BODY DIMENSIONS: 1 x 0.2 m (3.4 x 0.5 ft)
PAYLOAD: VHF transponder; UHF transponder

Sirius INTERNATIONAL

Communications

SPECIFICATION
(Sirius 3):

MANUFACTURER: Hughes Space
and Communications (now
Boeing)
LAUNCH DATE: October 5, 1998
ORBIT: 5°E (GEO)
LAUNCH SITE: Kourou, French
Guiana
LAUNCHER: Ariane 44L
LAUNCH MASS: 1465 kg (3223 lb)
BODY DIMENSIONS: 2.2 x 3.3 m
(7.1 x 10.9 ft)
PAYLOAD: 15 Ku-band
transponders

SES SIRIUS AB is owned by SES ASTRA and the Swedish Space Corporation. As such, it is part of SES, the world's largest network of satellite operators, with 44 satellites in its fleet. SES SIRIUS owns and operates two SIRIUS satellites that deliver TV, radio, data, Internet, and multimedia communications to the Nordic and Baltic regions and Eastern and central Europe.

SIRIUS 3 is a high-power satellite providing direct-to-home and cable television services as well as data distribution in Scandinavia and neighboring countries. It is also capable of providing television distribution and high-speed Internet data to Greenland. The three-axis-stabilized spacecraft is based on the HS 376 bus and uses gallium arsenide solar cells to generate a minimum of 1.4 kW of power at end of life. The Sirius 3 antenna has octagonal surface reflectors of approximately 2 m (6.6 ft) in diameter, with single offset feeds. This antenna has three surfaces: one for horizontally polarized signals, one for vertically polarized signals, and one for on-station tracking and command. Operational lifetime is 12 years.

Spaceway USA

Communications

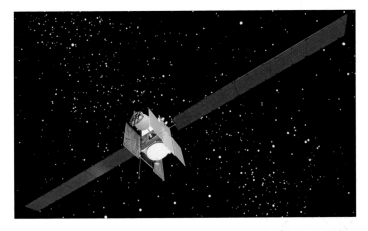

The first two Spaceway satellites are owned and operated by US company DIRECTV. They were originally commissioned as part of the Hughes Spaceway global broadband communications system. After the Hughes DTH business was taken over by DIRECTV, the satellites were retrofitted to deliver high-definition local TV channels across the USA.

Spaceway 1, launched in April 2005, was the heaviest commercial communications satellite launched by a Zenit 3SL. Spaceway 2, also a three-axis-stabilized Boeing 702 model satellite with a 12-year life expectancy, was moved to an Ariane 5 to ensure the satellites were both launched in 2005. It features onboard digital processors, packet switching, and spot-beam technology, providing high-speed, two-way communications for Internet, data, voice, video, and multimedia applications. The spacecraft includes a flexible payload with a fully steerable downlink antenna that can be reconfigured on orbit. Two solar arrays span 40.9 m (134.2 ft) and produce 12.3 kW at end of life.

SPECIFICATION (Spaceway 2):

MANUFACTURER: Boeing Satellite Systems
LAUNCH DATE: November 16, 2005
ORBIT: 99.2°W (GEO)
LAUNCH SITE: Kourou, French Guiana
LAUNCHER: Ariane 5 ECA
LAUNCH MASS: 6116 kg (13,455 lb)
BODY DIMENSIONS: 5.1 x 3.4 x 3.2 m (16.6 x 11.2 x 10.6 ft)
PAYLOAD: 72 Ka-band transponders

Superbird JAPAN

Communications

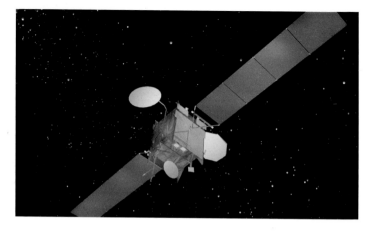

The Space Communications Corporation, a consortium of leading Japanese companies, was formed in 1985; its first communications satellite, Superbird-A, was launched four years later. Today, SCC operates four communications satellites named A, B2, C, and D at four orbital slots. Superbird-6 (A2), launched in 2004, was placed in a lower perigee than expected and suffered damage to a solar array. One of two main fuel tanks then lost pressure on November 28, causing an attitude error.

Superbird-D (also known as N-SAT-110 and JCSat-110) was built for use by SCC and Japan Satellite Systems (JSAT). It is based on the Lockheed Martin A2100AX satellite bus. The three-axis-stabilized spacecraft features high-output 120 W Ku-band transponders and a circular downlink. The solar arrays span 26.4 m (86.6 ft) and will provide 8.3 kW at end of life. It provides high-volume communications and broadcasting services to all of Japan and is used for the JSAT direct-to-home service of SKY PerfecTV!. Design life is 13 years.

SPECIFICATION
(Superbird-D):

MANUFACTURER: Lockheed Martin Commercial Space Systems
LAUNCH DATE: October 7, 2000
ORBIT: 110°E (GEO)
LAUNCH SITE: Kourou, French Guiana
LAUNCHER: Ariane 42L
LAUNCH MASS: 3530 kg (7766 lb)
BODY DIMENSIONS: Height, 6 m (19.7 ft)
PAYLOAD: 24 Ku-band transponders

TDRS (Tracking and Data Relay Satellite) USA

Communications

SPECIFICATION
(TDRS 10):

MANUFACTURER: Boeing Satellite
Systems
LAUNCH DATE: December 5, 2002
ORBIT: 40.9°W (GEO)
LAUNCH SITE: Cape Canaveral,
Florida
LAUNCHER: Atlas IIA
LAUNCH MASS: 3196 kg (7039 lb)
BODY DIMENSIONS: 3.4 x 3.4 x
8.4 m (11.7 x 11.7 x 27.7 ft)
PAYLOAD: S-band transponders;
Ku-band transponders; Ka-band
transponders; navigation
package

The Tracking and Data Relay Satellite System is the backbone of NASA's space-to-ground communications, providing almost uninterrupted voice, television, digital, and analog data communications with Earth-orbiting Space Shuttles and satellites. The first TDRS was launched from the Space Shuttle *Challenger* (STS-6) in April 1983. At that time it was the largest, most sophisticated communications satellite ever built. Its upper-stage motor failed and the satellite took six months to reach GEO. This reduced its operational lifetime, and it has since been reduced to part-time duty, such as supporting Antarctic communications. The second satellite was lost in the *Challenger* disaster and later replaced by TDRS 7.

The first seven satellites were built by TRW (now Northrop Grumman), while TDRS 8(H), 9(I), and 10(J) were built by Boeing, using the HS 601 bus. Each has two steerabl reflectors, 5 m (16.4 ft) in diameter, that can simultaneously transmit and receive on S-band and either Ku- or Ka-band. The newer satellites can receive data at 300 Mbps using Ku- and Ka-band, and 6 Mbps on S-band. Each spacecraft carries the additional capability for Ka-band receive rates of up to 800 Mbps.

Thaicom THAILAND

Communications

SPECIFICATION
(Thaicom-5):

MANUFACTURER: Alcatel Alenia
Space
LAUNCH DATE: May 27, 2006
ORBIT: 78.5°E (GEO)
LAUNCH SITE: Kourou, French
Guiana
LAUNCHER: Ariane 5 ECA
LAUNCH MASS: 2766 kg (6085 lb)
BODY DIMENSIONS: 3.7 x 3.3 x
2.2 m (12.1 x 10.8 x 7.2 ft)
PAYLOAD: 25 C-band
transponders; 14 Ku-band
transponders

Thailand inaugurated its first national GEO communications network in 1993–94 with the launches of Thaicom-1 and Thaicom-2, followed by Thaicom-3 in April 1997 and Thaicom-4 (IPSTAR) in August 2005. The Shinawatra Satellite Company of Bangkok operates the fleet under a lease arrangement with the Thai government.

The latest in the fleet is Thaicom-5, a three-axis-stabilized spacecraft based on the Alcatel Alenia Spacebus 3000A platform. Power output at end of life is 5 kW and design life is 12 years. Thaicom-5 provides telecom and television services for the entire Asia-Pacific region. Global beam coverage spans most of Asia, Europe, Australia, and Africa. The high-powered Ku-band transponders, with both spot and steerable beams, are ideally suited to digital direct-to-home services for Thailand and other countries in the region. It will eventually replace Thaicom-3.

Thor NORWAY

Communications

SPECIFICATION
(Thor 3):

MANUFACTURER: Hughes Space
and Communications (now
Boeing)
LAUNCH DATE: June 9, 1998
ORBIT: 0.8°W (GEO)
LAUNCH SITE: Cape Canaveral,
Florida
LAUNCHER: Delta II 7925
LAUNCH MASS: 1750 kg (3850 lb)
BODY DIMENSIONS: 3.3 x 2.2 m
(10.5 x 7.1 ft)
PAYLOAD: 14 Ku-band
transponders

Norway's Telenor Satellite Broadcasting is the largest provider of TV services
in the Nordic region. Telenor currently owns two Thor satellites and is co-
owner, with Intelsat, of the Intelsat 10-02 spacecraft. Both Thor satellites are
located at Telenor's 1°W orbital slot. Thor 2, which is to be retired by early
2009, will be replaced by Thor 2R, currently under construction by Orbital
Sciences Corporation.

Both Thor 2 and Thor 3 are based on HS 376 models. The cylindrical,
three-axis-stabilized satellites use gallium arsenide solar cells to generate
1.4 kW of power at end of life and rely on nickel hydrogen batteries for
power through eclipses. Thor 2 has 15 active Ku-band transponders (with
three spares), while Thor 3 has 14 active Ku-band transponders. Their
antennas have octagonal surface reflectors about 2 m (6.6 ft) in diameter,
with single offset feeds. These powerful satellites deliver direct-to-home
television to Scandinavia and northern Europe. Design life is about 12 years.
Thor 3 is expected to be retired in 2010.

Thuraya UNITED ARAB EMIRATES

Communications

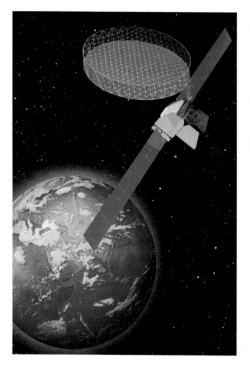

SPECIFICATION
(Thuraya-2):

MANUFACTURER: Boeing
LAUNCH DATE: July 10, 2003
ORBIT: 44°E (GEO)
LAUNCH SITE: Pacific Ocean
LAUNCHER: Zenit-3SL
LAUNCH MASS: 5250 kg
(11,578 lb)
BODY DIMENSIONS: 7.6 x 3.2 x
3.4 m (24.9 x 10.5 x 11.2 ft)
PAYLOAD: 128 L-band
transponders; 2 C-band
transponders

Thuraya was founded in the United Arab Emirates in 1997 by a consortium of national telecommunications operators and international investment houses. The concept was to offer cost-effective satellite-based mobile-telephone services to Europe, the Middle East, North and Central Africa, and Central and South Asia. Thuraya-1 was launched in October 2000, followed by a second satellite three years later. Thuraya-D3 will follow in late 2007.

The high-power Thuraya satellites are the first spacecraft in the Boeing GEM series. The satellites (derived from the Boeing 702 body-stabilized design) generate 11 kW and provide a range of voice and data services over a large geographic region, carrying many transponders to relay mobile telephone calls between countries in and around the Middle East and the Indian subcontinent. Thuraya's 200 spot beams can be steered to meet varying call densities, and enable it to handle 13,750 calls simultaneously. It carries a 12 m (39.4 ft) antenna for L-band mobile communications.

XM Radio USA

Digital radio communications

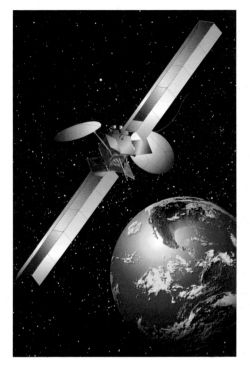

SPECIFICATION (XM-4):

MANUFACTURER: Boeing Satellite Systems
LAUNCH DATE: October 30, 2006
ORBIT: 115°W (GEO)
LAUNCH SITE: Odyssey platform, Pacific Ocean
LAUNCHER: Zenit 3SL
LAUNCH MASS: 5193 kg (11,425 lb)
BODY DIMENSIONS: 2.0 x 3.2 x 3.6 m (6.6 x 10.5 x 11.8 ft)
PAYLOAD: 2 S-band transponders

XM Satellite Radio owns and operates four spacecraft that provide digital radio programming directly to cars, homes, and portable radios coast-to-coast in the United States and parts of Canada. The first two XM satellites—XM-Rock and XM-Roll—were launched in 2001. XM-3 (XM-Rhythm) was launched in 2005, followed by XM-4 (also known as XM-Blues) in 2006.

Built by Boeing with a high-power S-band digital audio radio payload supplied by Alcatel Alenia Space, XM-4 will replace XM-l and -2, which are expected to reach the end of their lives in 2008 and will then become in-orbit spares. Two active transponders generating approximately 3 kW of power are the most powerful commercial transponders ever built. The Boeing BSS-702 model adds to XM Satellite Radio's in-orbit capacity to serve its 7 million radio subscribers. It is designed to deliver 18 kW of power at the beginning of its contracted 15-year service life. A fifth XM unit, now under construction at Space Systems/Loral, is scheduled to be completed by late 2007.

Military
Satellites

AEHF (Advanced Extremely High Frequency, Milstar 3)

USA

Communications

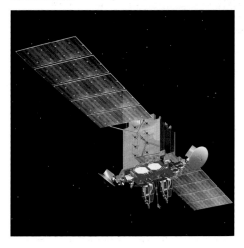

SPECIFICATION:

MANUFACTURER: Lockheed Martin Space Systems
LAUNCH DATE: 2008?
ORBIT: GEO
LAUNCH SITE: Cape Canaveral, California
LAUNCHER: Delta IV or Atlas V
LAUNCH MASS: Approx. 6136 kg (13,500 lb)
BODY DIMENSIONS: Length 9.75 m (32 ft)
PAYLOAD: Onboard signal processing; crosslinked EHF/SHF communications

AEHF will replace the Milstar 2 series, providing global, highly secure communications for US armed forces. They will form the backbone of the DOD's intermediate-term Military Satellite Communications (MILSATCOM) architecture until the larger-capacity Transformational Communications Satellite System or its equivalent enters service.

Based on Lockheed Martin's A2100 bus, the core structure contains the integrated propulsion system, while the payload module carries the electronics as well as the payload processing, routing, and control hardware and software that perform the satellite's communications function. Northrop Grumman Space Technology provides the payload. The AEHF system will have ten times the capacity and deliver data at six times the speed of the Milstar 2 satellites. The higher data rates permit transmission of tactical military communications such as real-time video, battlefield maps and targeting data.

Originally, five AEHF satellites were to be built, four crosslinked satellites providing coverage of the Earth from 65° N to 65° S plus one spare. In November 2002, two satellites were canceled and the third satellite was delayed. Lockheed Martin is now contracted to provide three satellites and the command control system for the Military Satellite Communications Systems Wing at Los Angeles Air Force Base.

Araks (Arkon) RUSSIA

Imagery intelligence

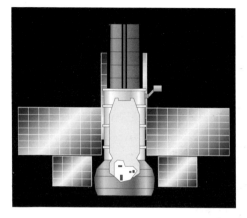

SPECIFICATION
(Kosmos-2392/Araks 2):

MANUFACTURER: NPO Lavochkin
LAUNCH DATE: July 25, 2002
ORBIT: 1506 x 1774 km (935 x 1102 mi), 63.5° inclination
LAUNCH SITE: Baikonur, Kazakhstan
LAUNCHER: Proton K/Block DM
LAUNCH MASS: 2600 kg (5720 lb)
PAYLOAD: CCD imaging system

First authorized in 1983, this third generation of military photo-reconnaissance satellites began with the development of a multimission platform based on the Arkon bus of NPO Lavochkin. The name used by Russia's space forces is Araks, after a river in the Caucasus. The original intention was to launch two groups of ten satellites into orbits at different altitudes.

Delayed by technical and budgetary problems, only two spacecraft have so far been launched. The first test flight was by Kosmos-2344, which was launched on June 6, 1997, into an unusually high (1516 x 2747 km (942 x 1707 mi) orbit that considerably reduced spatial resolution. It apparently failed after four months instead of the estimated two years. It was followed by the much smaller Kosmos 2392 in 2002. This satellite has also ceased to be operational.

The satellite is placed into a parking orbit by a Proton K and then moved to its operational orbit by two burns of the Block DM5 upper stage. A reflecting telescope system with a very long focal length gives ground resolution of 2–10 m (6.6–33 ft) with a typical swath width of 15–35 km (9.3–21.7 mi), depending on altitude. A CCD sensor operates in eight bands in the optical and near infrared. The use of infrared sensors, which operate in low temperatures, requires a large cooling system on the side of the spacecraft. The satellite can be pointed 20° from nadir, and has a rapid revisit capability.

COSMO-SkyMed (Constellation of Small Satellites for Mediterranean Basin Observation) ITALY

Radar imagery intelligence

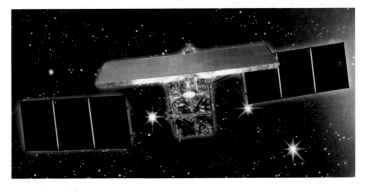

COSMO-SkyMed is a joint civil-military program funded by Agenzia Spaziale Italiana and the Italian Ministry of Defense. The system, now being built, consists of a constellation of four mid-sized satellites in LEO, each carrying a multimode, high-resolution X-band Synthetic Aperture Radar (SAR) with best resolution of less than 1 m (3.3 ft). The primary mission is to provide services for land monitoring, territory strategic surveillance, management of environmental resources, maritime and shoreline control, and law enforcement.

The satellite is based on the PRIMA (translates as Reconfigurable Italian Platform for Multiple Applications) bus of Alcatel Alenia Space (now Thales Alenia). The three-axis-stabilized spacecraft consists of the main bus, two solar arrays, and a 5.75 x 1.5 m (18.9 x 4.9 ft) SAR antenna on the nadir side.

All satellites will operate on the same orbital plane, following a dawn–dusk Sun-synchronous orbit with a 16-day repeat cycle. 3-D SAR imagery will be produced by combining two slightly different radar views of the same target. The series will be completed by 2009. Design life is five years.

SPECIFICATION (COSMO-SkyMed 1):

MANUFACTURER: Thales Alenia Space Italia
LAUNCH DATE: June 7, 2007
ORBIT: 620 km (385 mi), 97.86° inclination (sun-synchronous)
LAUNCH SITE: Vandenberg, California
LAUNCHER: Delta II 7420
LAUNCH MASS: Approx. 1700 kg (3740 kg)
PAYLOAD: Synthetic Aperture Radar (SAR)

Defense Meteorological Satellite Program (DMSP) USA

Meterology

SPECIFICATION
(DMSP F-17):

MANUFACTURER: Lockheed Martin
LAUNCH DATE: November 4, 2006
ORBIT: 846 x 850 km (526 x 528 mi), 98.8° inclination (Sun-synchronous)
LAUNCH MASS: 1220 kg (2684 lb)
BODY DIMENSIONS: Approx. 1.2 x 6.4 m (3.9 x 21 ft) deployed
LAUNCH SITE: Vandenberg, California
LAUNCHER: Delta IV-Medium
PAYLOAD: Operational Linescan System (OLS) imager; Special Sensor Microwave Imager/ Sounder (SSMIS); Special Sensor Precipitating Electron Spectrometer (SSJ5); Special Sensor Ion/Electron Scintillator (SSIES3); Special Sensor Magnetometer (SSM); Laser threat warning sensor

Since 1965, 44 Lockheed Martin DMSP satellites have been launched by the US Air Force to aid US forces in planning operations. The satellites are operated by National Oceanic and Atmospheric Administration and provide global weather data for US armed services. The Air Force Weather Agency at Offutt Air Force Base in Omaha, Nebraska, is the primary data user within the US. The Space and Missile Systems Center at Los Angeles Air Force Base, California, manages the program.

DMSP F-17 is the second Block 5D-3 spacecraft built under contract for the USAF by Lockheed Martin. Placed in a morning, Sun-synchronous orbit, it replaced DMSP 5D-2 F-13, which was launched in 1995. The Block 5D-3 series carries larger sensor payloads than earlier generations. They also feature a more capable power subsystem, a more powerful onboard computer with increased memory, and increased battery capacity. Starting with F-17, the attitude-control subsystem has also been enhanced by the integration of a second inertial measurement unit using ring laser, instead of mechanical, gyros to provide greater precision pointing flexibility.

Visible and infrared sensors image cloud cover and measure precipitation, surface temperature, and soil moisture. There are always two operational spacecraft in near-polar orbits. After DMSP F-17, three satellites remain to be launched.

Defense Satellite Communications System (DSCS) USA

Communications

SPECIFICATION (DSCS 3-B6):

MANUFACTURER: Lockheed Martin Astro Space
LAUNCH DATE: August 29, 2003
ORBIT: 85° W (GEO)
LAUNCH SITE: Cape Canaveral, Florida
LAUNCHER: Delta 4-Medium
LAUNCH MASS: 1232 kg (2716 lb)
DIMENSIONS: 2 x 1.9 x 1.9 m (6.6 x 6.2 x 6.2 ft)
PAYLOAD: 4 C-band transponders

The DSCS constellation, which includes nine active satellites in geostationary orbit, is the US Air Force's longest-running communications satellite system. It provides uninterrupted, secure voice and high-data-rate communications to the DOD and other government users, including troops and military commanders at multiple locations.

The first generation of DSCS satellites, weighing about 45 kg (99 lb), were launched between 1966 and 1968. The DSCS 2 program began in 1971, expanding the services provided by their predecessors. The first of the third generation was launched in 1982, overlapping the DSCS 2 series.

The last satellite, DSCS 3-B6, was scheduled to launch with DSCS 3-A3 aboard a 1986 DOD Space Shuttle mission. After the *Challenger* disaster, the satellites were significantly modified, with major enhancements to their capabilities. The A3 satellite eventually flew in 2003, followed by B6. This brought the total to 14 DSCS satellites in orbit, of which nine are active. The final four spacecraft have service-life-enhancement upgrades that provide increased downlink power, improved connectivity to their antennas, and upgraded transponder channels. Design life is 10 years. Their successor is the Wideband Global SATCOM.

Defense Support Program (DSP) USA

Ballistic missile early warning

The DSP satellites, operated by the USAF Space Command, provide early warning of missile launches, space launches, and nuclear detonations. Originally built by TRW, the first Phase I satellite was launched in November 1970. It weighed about 900 kg (2000 lb), with 400 W of power, 2000 detectors, and a design life of 1 year 3 months. The current version, the DSP-1 series (satellites 14 to 23), has 1485 W of power, 6000 detectors, and a design life of at least 10 years.

The satellites use infrared sensors to detect heat from missile plumes against the Earth's background. With up to 10 satellites still operational, they can provide stereo views of launches and better plume characterization. Recent improvements in sensor design include above-the-horizon capability for full hemispheric coverage and improved resolution, and increased onboard signal-processing capability.

Most DSP satellites were launched into GEO by a Titan IV. DSP-16 was launched from Shuttle *Discovery* on mission STS-44 (November 24, 1991). The twenty-third and final satellite will be launched directly into GEO by a Delta 4-Heavy. DSP will be replaced by the Space-Based Infrared System.

SPECIFICATION (DSP-23):

MANUFACTURER: Northrop-Grumman
LAUNCH DATE: November 10, 2007
ORBIT: GEO
LAUNCH SITE: Cape Canaveral, Florida
LAUNCHER: Delta IV-Heavy
LAUNCH MASS: 2386 kg (5250 lb)
BODY DIMENSIONS: 10 x 6.7 m (32.8 x 13.7 ft)
PAYLOAD: Medium-infrared wavelength telescope; mercury cadmium telluride detector; dual radiation detectors; Space Atmosphere Burst Reporting System (SABRS)

Global Positioning System (GPS) USA

Navigation

SPECIFICATION
(Navstar-2RM3/Block
IIR-16M):

MANUFACTURER: Lockheed Martin
Space Systems
LAUNCH DATE: November 17,
2006
ORBIT: 20,182 km (12,540 mi),
55° inclination (plane B4)
LAUNCH SITE: Cape Canaveral,
Florida
LAUNCHER: Delta II 7925
LAUNCH MASS: 2032 kg (4479 lb)
BODY DIMENSIONS: 1.5 x 1.9 x
1.9 m (5 x 6.3 x 6.3 ft)
PAYLOAD: L-band and S-band
transmitters

The Navstar GPS program is a dual-use, satellite-based system that provides accurate positioning, navigation, and timing information to military and civilian users worldwide. Originally developed by the US DOD to provide precise positional information and weapon guidance, the system now supports a wide range of civil, scientific, and commercial functions. GPS is based on 24 satellites in six orbit planes, with four satellites per plane. Each is fitted with four atomic clocks that emit timing signals. When several signals are detected, the location of the receivers can be accurately determined.

After a number of preliminary test flights, a GPS technology prototype, Navstar 1, was launched on February 22, 1978. Eleven of the Block I satellites were launched between 1978 and 1985. They were replaced by the operational Block 2 and 2A series, with 28 launches between 1989 and 1997. GPS Block IIR satellites were introduced in 1997, carrying improved clocks and reprogrammable processors to enable fixes and upgrades in flight. The GPS Block IIR-M (Modernized) satellites were introduced in September 2005. Improvements included a new military signal on the L1 and L2 channels, and a more robust civil signal. In September 2007, the constellation consisted of 16 Block IIA satellites built by Boeing and 12 Block IIR and three Block IIR-M satellites built by Lockheed Martin, with five GPS Block IIR-M satellites remaining to be launched. Block IIF is under development.

Glonass (Global Navigation Satellite System) RUSSIA

Navigation

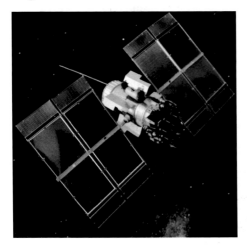

SPECIFICATION (Kosmos 2424):

MANUFACTURER: NPO-PM, NPO Reshetnev
LAUNCH DATE: December 24, 2006
ORBIT: 19,100 km (11,868 mi), 64.8° inclination
LAUNCH SITE: Baikonur, Kazakhstan
LAUNCHER: Proton K
LAUNCH MASS: 1415 kg (3113 lb)
BODY DIMENSIONS: 2.4 x 3.7 m (7.9 x 12.1 ft)
PAYLOAD: L-band transmitters; Ku-band transponders

Glonass is a radio-based satellite navigation system developed by the Soviet Union and now operated for the Russian government by the national space forces. It is the Russian counterpart of the US Global Positioning System (GPS). Today, Glonass has both military and civilian applications, but funding problems mean that the constellation remains incomplete. A fully operational system consists of 24 satellites: three orbital planes, each with eight spacecraft, including three in-orbit spares. Russian officials have declared their intention to make Glonass comparable in precision to the US GPS by 2011.

The Glonass program began with three satellites (Kosmos 1413–15), launched by a Proton on October 12, 1982. They were inserted into a low parking orbit and then moved to their operational orbit after two burns of the upper stage. Ten Block I satellites were launched but average lifetime was only 14 months.

The latest version is the Glonass-M or Uragan-M, first launched on December 10, 2003. It has two solar arrays providing 1.4 kW of power, a longer service life of seven years, updated antenna feeder systems, and an additional navigation frequency for civilian use. Kosmos 2424–26 were launched to repopulate plane 2; planes 1 and 3 had operational satellites. In October 2007, the constellation consisted of nine operational spacecraft with one being commissioned. A third generation, Glonass-K, is under development.

Helios FRANCE

Imagery intelligence

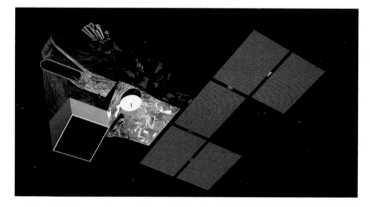

Helios was Europe's first family of military optical surveillance satellites, based on the SPOT spacecraft bus and imaging system. The first generation, Helios 1A and Helios 1B, were launched on July 7, 1995, and December 3, 1999. The first remains operational, but 1B's mission ended when its four batteries failed in October 2004.

The third satellite, Helios 2A, is a collaborative effort with Spain and Belgium. This is a more advanced version derived from the SPOT 5 platform, with greatly improved maneuverability. A single solar array provides 2.9 kW of power. The new-generation Helios 2 offer spatial resolution of about 0.5 m (1.6 ft), infrared capability (for night observation), greater imaging capacity with simultaneous imaging in several modes, and shorter image-transmission times.

Helios carries a VHR imaging system operating in the visible and infrared and a WA instrument that is similar to the SPOT 5 system. In addition to a day/night intelligence capability, Helios 2 is used for targeting, guidance, mission planning, and combat damage verification. Helios 2B is expected to be launched in 2008. Design life is five years.

SPECIFICATION (Helios 2A):

MANUFACTURER: EADS-Astrium
LAUNCH DATE: December 18, 2004
ORBIT: 690 km (429 mi), 98.08° inclination (Sun-synchronous)
LAUNCH SITE: Kourou, French Guiana
LAUNCHER: Ariane 5G
LAUNCH MASS: 4200 kg (9240 lb)
BODY DIMENSIONS: 6 x 3.7 x 3.4 m (19.7 x 12.1 x 11.2 ft)
PAYLOAD: Very High Resolution (VHR) imaging system; Wide Angle (WA) imaging system

Midcourse Space Experiment (MSX) USA

Experimental ballistic missile warning satellite

SPECIFICATION:

MANUFACTURER: Johns Hopkins
University Applied Physics Lab
LAUNCH DATE: April 24, 1996
ORBIT: 908 km, 99.6° inclination
(near Sun-synchronous)
LAUNCH SITE: Vandenberg,
California
LAUNCHER: Delta II 7920
LAUNCH MASS: 2700 kg (5940 lb)
BODY DIMENSIONS: 5 x 1.5 x
1.5 m (16.4 x 4.9 x 4.9 ft)
PAYLOAD: Space Infrared
Imaging Telescope (SPIRIT III);
Ultraviolet and Visible Imagers
and Spectrographic Imagers
(UVISI); Space-Based Visible
sensor (SBV); Onboard Signal
Data Processor (OSDP);
contamination sensors; mirror
cleaning experiment

MSX was developed by the US Ballistic Missile Defense Agency to characterize ballistic missile signatures during their midcourse phase of flight and to collect data on the backgrounds against which targets are seen. It also carried out scientific investigations of the celestial background and the Earth's atmosphere, observing ozone, chlorofluorocarbons, carbon dioxide, and methane.

MSX comprises an instrument section, a truss structure, and an electronics section. Instruments account for about half of its total mass. The instrument section houses 11 optical sensors, which are precisely aligned so that target activity can be viewed simultaneously by multiple sensors. The MSX sensors are the first hyperspectral imagers flown in space and cover the spectrum from the far ultraviolet to the very-long-wave infrared.

The infrared sensors were cooled to 11–12 K by a solid hydrogen cryostat. The cryogen mission phase ended on February 26, 1997, but the other sensors remained operational. In July 1999, MSX was used to observe shuttle thruster firings on STS-93. Air Force Space Command took over ownership in October 2000. Today, the Space-Based Visible Sensor, operated by MIT, is the only sensor in use. It is the command's only space-based surveillance instrument able to provide full metric and identification coverage of objects in geosynchronous orbit around the Earth.

Milstar USA

Communications

SPECIFICATION
(Milstar 6):

MANUFACTURER: Lockheed Martin
LAUNCH DATE: April 8, 2003
ORBIT: 89.8° W (GEO)
LAUNCH SITE: Cape Canaveral,
Florida
LAUNCHER: Titan IV/Centaur
LAUNCH MASS: 4536 kg (9979 lb)
BODY DIMENSIONS: 15.5 m
(50.7 ft) length
PAYLOAD: Low data rate (LDR)
communications system;
Medium data rate (MDR)
communications system

Milstar provides the US DOD and operational forces with secure, jam-proof communications between fixed, mobile, and portable terminals. It is the first satellite communications system to use signal-processing algorithms on the satellites, allowing service commanders to establish customized networks within minutes. The satellites are connected in a ring configuration encircling the Earth. The first Milstar was launched by a Titan IV on February 7, 1994. Since then, five more have been launched, though Milstar 3 was placed in the wrong orbit and is nonoperational. The current constellation comprises five operational spacecraft. Two of these are the first-generation Block I design, with an LDR payload built by TRW and two crosslink antennas built by Boeing Satellite Systems.

The first successful launch of the upgraded Milstar II was in February 2001. The Block II series (Milstar 3–6) offer added security through specially designed antennas and faster data-rate transmissions via the Boeing-built MDR payload that has 32 channels and can process data at 1.5 Mbits per second. The MDR also includes two nulling spot antennas that can identify and pinpoint the location of a jammer and electronically isolate its signal. Two solar arrays provide 5 kW of power. The last Milstar was launched in April 2003. Design life is 10 years. Milstar will be replaced by AEHF.

Molniya RUSSIA

Civil and military communications

SPECIFICATION
(Molniya 3-53):

MANUFACTURER: NPO-PM/
NPO Reshetnev
LAUNCH DATE: June 19, 2003
ORBIT: 651 x 39,703 km
(404 x 24,670 mi), 62.9°
inclination
LAUNCH SITE: Plesetsk, Russia
LAUNCHER: Molniya
LAUNCH MASS: 1740 kg (3828 lb)
BODY DIMENSIONS: 1.6 x 4.2 m
(5.2 x 13.8 ft)
PAYLOAD: Three 4–6 GHz
transponders

Molniya (Lightning) was the first Soviet communications satellite network. It was designed to provide communications for the northern regions of the huge country from highly inclined, elliptical orbits. However, the importance of these satellites decreased as more GEO Comsats were launched.

The satellites are based on the cylindrical KAUR-2 bus, with a domed end and six solar arrays in windmill configuration, providing 1.2 kW of power. The first successful Molniya launch took place on April 23, 1965, relaying TV transmissions and two-way telephone and telegraph communications. One transponder was probably dedicated for military/government use.

Molniya 1T was used almost exclusively for military communications. With Molniya 1-11 in 1969, the number of satellites above the horizon at any time was increased from two to three by reducing the equatorial spacing to 90°.

The Molniya 3 series first flew in November 1974 and was used to create the Orbita communications system for northern regions of the USSR. The Molniya 1 and 3 satellites were placed in eight orbital planes at 45° intervals. Fifty-five Molniya 3 satellites have been launched. Design life is three years. The latest version is the Molniya 3K, first launched on July 20, 2001. After a launch failure in July 2005, it was announced that Molniya 3K will cease production.

Ofeq (Horizon)—also known as Ofek—is Israel's home-grown optical reconnaissance satellite program. The first Ofeq was launched on September 19, 1988. Since then, there have been five successful launches and two failures. The satellites are launched towards the west to avoid dropping stages on Israel's Arab neighbors. The east-to-west orbit at 38° inclination is phased to give optimal daylight coverage of the Middle East.

Ofeq is based on the same IAI bus as the EROS civil remote sensing satellite. The lightweight, three-axis-stabilized spacecraft has a rapid body-pointing capability of 45° from nadir in all directions. It carries two solar arrays and improved digital cameras, produced by Israel's Elbit Systems, that provide ground resolution of 0.4 m (1.3 ft), compared with 0.8 m (2.6 ft) for Ofeq 5. The Ofeq 6 satellite was lost in a Shavit launch failure in September 2004. The latest addition, Ofeq 7, ensured continuation of the program alongside the five-year-old Ofeq 5. Design life is four years.

SPECIFICATION (Ofeq 7):

MANUFACTURER: Israel Aerospace Industries
LAUNCH DATE: June 10, 2007
ORBIT: 340 x 575 km (211 x 357 mi), 141.8° inclination
LAUNCH SITE: Palmachim, Israel
LAUNCHER: Shavit-2
LAUNCH MASS: 350 kg (770 lb)
BODY DIMENSIONS: 1.2 x 3.6 m (3.9 x 11.8 ft)
PAYLOAD: Panchromatic imaging camera

Oko (US-KS) RUSSIA

Ballistic missile early warning

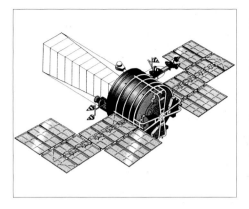

SPECIFICATION (Kosmos-2422):

MANUFACTURER: NPO Lavochkin
LAUNCH DATE: July 21, 2006
ORBIT: 539 x 39,570 km
(335 x 24,588 mi), 62.9°
inclination (semi-synchronous)
LAUNCH SITE: Plesetsk, Russia
LAUNCHER: Molniya-M
LAUNCH MASS: Approx. 2400 kg
(5280 lb)
BODY DIMENSIONS: 2 x 1.7 m
(6.6 x 5.6 ft)
PAYLOAD: 50 cm (19.7 in)
telescope; visible/infrared
sensors

Oko (Eye) is a series of Russian missile early warning satellites, with some similarities to the American DSP program. This system was built to detect launches of ballistic missiles from the US and Western Europe and cannot detect missiles launched from sea or other regions. The first test launch (Kosmos 520) took place in September 1972, but the system only became fully operational in 1987, after many years of development. Only seven of the first 13 satellites worked more than 100 days. Until 1983, they carried a self-destruct system that was activated if the satellite lost communication with ground control.

Oko consists of three main sections: an engine block, a payload section, and an optical section. It has a cylindrical bus with two solar arrays. There are four liquid-fuel orbit-correction engines and 16 liquid-fuel engines for orientation and stabilization. The first-generation Oko has a 50 cm (19.7 in) diameter telescope and an infrared solid-state sensor to detect heat from missiles. Several smaller telescopes provide a wide-angle view of the Earth in infrared and visible parts of spectrum. Images are sent to the ground in real time.

The Okos follow overlapping, highly elliptical orbits with apogees over Western Europe and the Pacific. Each views ICBM sites in the US for 5–6 h per revolution (12 h daily). A full system requires nine satellites operating 160 min apart. By September 2007, only Kosmos-2422 was operational.

Orbital Express (ASTRO and NextSat) USA

Autonomous rendezvous and docking

Orbital Express was part of a Defense Advanced Research Projects Agency (DARPA) program to demonstrate fully autonomous on-orbit satellite servicing capabilities. Using autonomous software provided by NASA-Marshall, its main objectives were to demonstrate fully autonomous rendezvous out to 7 km (4.3 mi); soft capture and autonomous station-keeping at submeter range; and on-orbit refueling and component replacement. Orbital Express comprised two satellites: the Autonomous Space Transport Robotic Operations (ASTRO) service vehicle, and the Next-generation Serviceable Satellite (NextSat). They were launched together on an Atlas V.

Despite problems with ASTRO's attitude-control system and flight software, Orbital Express completed several hydrazine transfers and robotic-arm demonstrations, including transfers of the battery orbital replacement unit using ASTRO's robotic arm. ASTRO was able to move up to 6 km (3.7 mi) from NextSat and then reestablish tracking using its infrared camera and laser range-finder at 3 km (1.9 mi). The final demonstration, which began on July 16, involved separating the satellites so that ASTRO lost sensor contact. With input from the ground-based Space Surveillance Network, it was able to approach NextSat again, switching to its onboard sensors to complete the rendezvous.

Mission lifetime was expected to be three months, but operations continued for four and a half months. On July 21, NextSat's solar arrays were turned away from the Sun and the satellite shut down. ASTRO's surplus fuel was dumped soon after, and it also was decommissioned.

SPECIFICATION (ASTRO):

MANUFACTURER: Boeing Phantom Works

LAUNCH DATE: March 8, 2007

ORBIT: 490 x 498 km (304 x 309 mi), 46° inclination

LAUNCH SITE: Cape Canaveral, Florida

LAUNCHER: Atlas V 401

LAUNCH MASS: 1091 kg (2400 lb)

BODY DIMENSIONS: 1.8 x 1.8 m (5.8 x 5.8 ft)

PAYLOAD: Autonomous rendezvous and capture sensors; capture subsystem; manipulator subsystem

SPECIFICATION (NextSat, shown below):

MANUFACTURER: Ball Aerospace & Technologies

LAUNCH DATE: March 8, 2007

ORBIT: 490 x 498 km (304 x 309 mi), 46° inclination

LAUNCH SITE: Cape Canaveral, Florida

LAUNCHER: Atlas V 401

LAUNCH MASS: 227 kg (500 lb)

BODY DIMENSIONS: 1 x 1 m (3.3 x 3.3 ft)

PAYLOAD: Passive capture mechanism; sensor system targets; cross-link communication; fluid transfer subsystem

Parus (Tsikada-M) RUSSIA

Navigation/store-dump communications

SPECIFICATION:
(Kosmos-2429)

MANUFACTURER: NPO-PM, NPO Reshetnev
LAUNCH DATE: September 11, 2007
ORBIT: 943 x 1008 km (586 x 626 mi), 83° inclination
LAUNCH SITE: Plesetsk, Russia
LAUNCHER: Kosmos-3M
LAUNCH MASS: 825 kg (1818 lb)
BODY DIMENSIONS: 2 x 3 m (6.7 x 9.8 ft)
PAYLOAD: VHF navigation system; store and forward communications relay system

Parus (Sail), sometimes referred to as Tsikada Military (Tsikada-M), provides specialized naviga-tion data and store-dump radio communications for Russian naval forces and ballistic missile sub-marines. The first test flights began with Kosmos-700 in 1974, and the system became operational in 1976. One satellite is launched each year.

The cylindrical Parus spacecraft uses the KAUR-1 bus and is almost identical to the Tsikada/Nadezdha used for civil navigation. The exterior is coated with solar cells providing about 200 W of power. The payload and telemetry antennas are attached to the bottom (Earth-facing) side. Each satellite transmits timing signals at two frequencies, enabling a receiving station to determine its location from the Doppler shift of the signal frequency. The constellation has at least six spacecraft in orbital planes 30° apart. There is a unique frequency for each plane and frequencies are different from the civil system. The high-inclination orbits provide global navigation support, although there is often no satellite above the horizon and long gaps in coverage.

SAR-Lupe GERMANY

Radar imagery intelligence

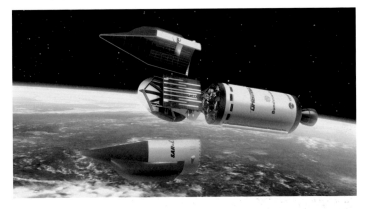

SAR-Lupe is a radar reconnaissance system for use by the German Federal Armed Forces. Germany's first satellite reconnaissance system, it consists of five identical satellites, to be launched at intervals of six months. The spacecraft fly in three different orbital planes, which enables them to observe the Earth's surface between 80°N and 80°S.

SAR-Lupe has a fixed, dual-use, parabolic antenna for radar observations and communications, and a side-mounted solar array. In high-resolution mode, the synthetic aperture radar (SAR) can detect objects as small as 50 cm (19.7 in) in diameter. It can also scan larger swaths at lower resolutions, providing a repetitive worldwide reconnaissance capability in all conditions. SAR data are transmitted on X-band frequencies, and encrypted S-band is used for telemetry and command transmissions direct to ground stations.

The first launch was in December 2006, using a Kosmos-3M with modified fairings to accommodate the large antennas. The second satellite followed on July 2, 2007. The project sponsors are the German Ministry of Defense and the Federal Office of Defense Technology and Procurement.

SPECIFICATION:

MANUFACTURER: OHB-System
LAUNCH DATE: December 19, 2006
ORBIT: 468 x 505 km (291 x 314 mi), 98.2° inclination (3 planes)
LAUNCH MASS: 770 kg (1694 lb)
BODY DIMENSIONS: Approx. 4 x 3 x 2 m (13.1 x 9.8 x 6.6 ft)
LAUNCH SITE: Plesetsk, Russia
LAUNCHER: Kosmos-3M
PAYLOAD: Synthetic Aperture Radar (SAR); intersatellite S-band transmitter; X-band transmitter

SBIRS (Space-Based Infrared System) USA

Ballistic missile early warning

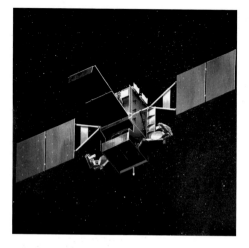

SPECIFICATION (SBIRS GEO-1):

MANUFACTURER: Lockheed Martin Space Systems
LAUNCH DATE: 2009?
ORBIT: GEO
LAUNCH SITE: Cape Canaveral, Florida
LAUNCHER: Atlas V?
LAUNCH MASS: 4900 kg (10,780 lb)?
PAYLOAD: Scanning infrared sensor; staring infrared sensor

Developed for the US Air Force Space and Missile Systems Center, SBIRS is the replacement for the US DSP early-warning satellites. It was originally intended to comprise satellites in both LEO and GEO. SBIRS-Low was later renamed the Space Tracking and Surveillance System (STSS). SBIRS-High remained simply SBIRS. Technical and funding issues have caused delays and reluctance to go ahead with the full system.

The SBIRS spacecraft is based on Lockheed Martin's A2100 bus. Up to four SBIRS satellites are planned for GEO, together with sensors carried as secondary payloads on two classified NRO satellites in highly elliptical orbits (HEO). From HEO, the infrared sensor will be able to detect missile and rocket launches at high latitudes.

The first HEO sensor (HEO-1) was launched aboard a classified NRO satellite (USA 184) on June 28, 2006, by a Delta IV from Vandenberg. Testing of the sensor and related systems should be completed in 2008, before the launch of the second HEO payload. The GEO payloads feature a scanning sensor that enables short revisit times over its full field of view and a staring sensor that can be tasked for step-stare or dedicated-stare operations over smaller areas. Communications from the infrared payload are protected by an antijamming system.

Sicral ITALY

Communications

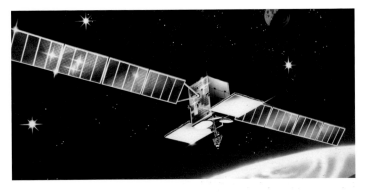

Sicral 1 is Italy's first military satellite and part of a three-nation contribution to provide telecommunications to NATO. It is designed to operate for 10 years, providing communications to fixed and mobile terminals operated by Italian and allied forces, including fighter aircraft. Sicral (Sistema Italiano per Comunicazioni Riservate ed Allarmi) is based on the Alenia Spazio (now Thales Alenia) GeoBus/Italsat bus. It is Europe's first spacecraft to use extremely high frequency (EHF) broadcast frequencies. The three-axis-stabilized spacecraft carries two solar arrays that provide 3.28 kW of power. Nine transponders transmit in the superhigh frequency, ultrahigh frequency and EHF bands, with the capability to switch bands onboard the satellite—a world first—to promote secure communications. The satellite has interoperability with US and other European satellite systems. Design life is 10 years.

The Italian Defense Ministry lost control of Sicral 1 in October 2006—possibly due to solar activity—and was forced to allow it to drift eastward before it returned to its orbital position in December. Sicral 1B is due to be launched in fall 2008.

SPECIFICATION (Sicral 1):

MANUFACTURER: Alenia Spazio, FiatAvio, Telespazio
LAUNCH DATE: February 7, 2001
ORBIT: 16.2° E (GEO)
LAUNCH SITE: Kourou, French Guiana
LAUNCHER: Ariane 44L
LAUNCH MASS: 2596 kg (5711 lb)
BODY DIMENSIONS: 2.8 x 3.4 x 4.9 m (9.2 x 11.3 x 16.2 ft)
PAYLOAD: 9 SHF, UHF, and EHF band transponders

Skynet 5 UK

Communications

SPECIFICATION
(Skynet 5A):

MANUFACTURER: EADS Astrium
LAUNCH DATE: March 11, 2007
ORBIT: 1° W (GEO)
LAUNCH SITE: Kourou, French Guiana
LAUNCHER: Ariane 5 ECA
LAUNCH MASS: 4635 kg (10,218 lb)
BODY DIMENSIONS: 4.5 x 2.9 x 3.7 m (14.8 x 9.5 x 12.1 ft)
PAYLOAD: 15 SHF-band channels; 9 UHF-band channels

Skynet 5 is the successor to the UK Defense Ministry's Skynet 4 constellation. It is used by the UK and NATO allies for secure, reliable military and government communications. Astrium Satellites is prime contractor for three Skynet 5 satellites through its subsidiary, Paradigm Secure Communications. Paradigm holds a £3.6 billion (USD 7.4 billion) contract with the UK Ministry of Defense for the provision of secure milsatcoms until 2020.

Skynet 5 offers a significant increase in capacity and performance over Skynet 4. It features nulling antennas for increased resistance to jamming, and hardening against nuclear radiation and laser attack. The three-axis-stabilized satellite is based on Astrium's high-power Eurostar E3000 platform, with a 34 m (111.5 ft) solar array span that provides 6 kW of power.

The Communication Module carries two helicoidal UHF antennas and a central S-band telemetry and telecommand mast antenna. Skynet 5A has multiple steerable spot beams and was the most powerful X-band satellite ever placed in orbit. A liquid propellant apogee engine (LAE) is used for orbit circularization. Its design lifetime is 15 years. At least two more in the series are to be launched in 2007–8. The last of the Skynet 4 satellites will be placed in a graveyard orbit by 2013–14.

Space Tracking and Surveillance System (STSS, SBIRS-Low) USA

Ballistic missile early warning

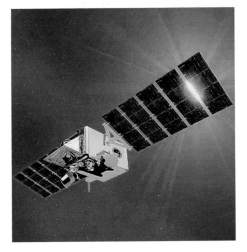

SPECIFICATION:

MANUFACTURER: Northrop Grumman
LAUNCH DATE: 2008?
ORBIT: LEO
LAUNCH SITE: Cape Canaveral, Florida
LAUNCHER: Delta II 7920
PAYLOAD: Infrared acquisition sensor; missile track sensor

STSS was previously known as Brilliant Eyes and then as SBIRS-Low. In 2002, the program was restructured by the US Missile Defense Agency and renamed. STSS is designed as a constellation of satellites in LEO, equipped with infrared and visible sensors that can acquire and track ballistic missiles.

Under the current plan, two Block 2006 research and development satellites will be launched into LEO by a single Delta II launch vehicle. They will carry two sensors, one for acquisition and another for tracking. They will test the key functions of a space-based sensor, passing short- and long-range ballistic-missile tracking data to ground operators to enable intercepts of missile targets. STSS's sensors will be able to track and discriminate missiles in midcourse; report on post-boost vehicle maneuvers, reentry-vehicle deployments, and the use of various types of decoys; and provide hit/kill assessment. Design life is four years.

The data obtained from these tests will be fed into the design of the second-generation operational satellites, which are not expected to be launched until 2016. A network of 30 advanced STSSs would be able to simultaneously detect and track more than 100 objects in real time and differentiate missiles and warheads from decoys, debris, clutter, and noise. They will carry two sensors, one with a wide-field view to capture the launch phase and a cryogenically cooled narrow-field sensor for tracking purposes.

Strela RUSSIA

Communications

SPECIFICATION
(Strela-3):

MANUFACTURER: NPO-PM
LAUNCH DATE: January 15, 1985
ORBIT: 1440 x 1450 km (894.8 x
901 mi), 82.5° inclination
LAUNCH SITE: Plesetsk, Russia
LAUNCHER: Tsyklon-3 or
Kosmos-3M
LAUNCH MASS: 225 kg (495 lb)
BODY DIMENSIONS: 1.5 x 1 m
(4.9 x 3.3 ft)
PAYLOAD: VHF/UHF data-relay
system

Strela (Arrow) is the name given to a series of military store-dump satellites used to relay messages to mobile users and the intelligence community. The system was developed in the 1960s, with 26 flights of the Strela-1 from August 18, 1964 to September 18, 1965. Strela-1 provided VHF/UHF medium-range links between armed forces. These 1 m (3.3 ft) diameter, 61 kg (134 lb) satellites were launched in groups of eight by Kosmos from Plesetsk into 1500 km (932 mi) circular orbits at 74° inclination.

The Strela-2 series conducted longer-range communications using three 875 kg (1925 lb) satellites built by AKO Polyot for NPO-PM. These flew in 786–810 km (488–503 mi) orbits at 74° inclination, spaced 120° apart.

The Strela-3 series began operations in 1985, when Kosmos 1617–1622 were launched by Tsyklon-3 from Plesetsk. Twelve spacecraft made up the operational constellation. The cylindrical spacecraft is stabilized using a gravity-gradient boom. It has 12 MB of onboard storage with a transmission rate of 2.4 kB/sec. Since 2002, when the Russian armed forces stopped using the Ukrainian-developed Tsyklon, Strela has been launched in pairs by the smaller Kosmos-3M. Some reports suggest that the most recent launch, Kosmos-2416 on December 21, 2005, involved a new-generation Strela replacement known as Rodnik.

Syracuse FRANCE

Communications

SPECIFICATION
(Syracuse 3B):

MANUFACTURER: Alcatel Alenia
Space (now Thales Alenia)
LAUNCH DATE: August 11, 2006
ORBIT: 5° W (GEO)
LAUNCH SITE: Kourou, French
Guiana
LAUNCHER: Ariane 5 ECA
LAUNCH MASS: 3750 kg (8250 lb)
BODY DIMENSIONS: 4 x 2.3 x
2.3 m (13.1 x 7.5 x 7.5 ft)
PAYLOAD: 9 SHF and 6 EHF
channels

Syracuse is a French advanced system for defense communications that ensures global voice and secure data links. It was originally developed in a collaboration between the French Ministry of Defense and CNES. Syracuse 1 and 2 consisted of a military communications band onboard seven French civilian Telecom 1 and 2 satellites. Syracuse 1 became operational in 1984, followed by Syracuse 2 in 1987. The Syracuse 3 series of dedicated satellites is designed to deliver higher data throughput, operational flexibility, and resistance to countermeasures and attack. Based on Alcatel's Spacebus platform, they are hardened to resist nuclear attack in compliance with NATO specifications. Their communications payload operates in two frequency bands. The SHF band has four spot beams, one global beam, and one beam for metropolitan France. The EHF band has two spot beams and one global beam. The steerable spot beams and global beams provide a large selection of footprints.

Syracuse 3A was launched on October 13, 2005, and was followed by Syracuse 3B in 2006. Syracuse 3A operates from 47° E. NATO chose the Syracuse 3 system for its Satcom Post 2000 project. This project aims to pool the Syracuse, Skynet, and Sicral satellite resources, so they can be shared with NATO allies. Design life is 12 years.

UHF Follow-On (UFO) USA

Communications

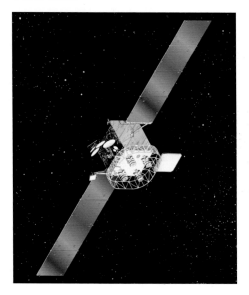

SPECIFICATION
(UHF-F11):

MANUFACTURER: Boeing
LAUNCH DATE: December 17,
2003
ORBIT: 172° E (GEO)
LAUNCH SITE: Cape Canaveral,
Florida
LAUNCHER: Atlas IIIB
LAUNCH MASS: Approx. 1364 kg
(3000 lb)
BODY DIMENSIONS: 3.4 x 3.2 x
3.4 m (11 x 10.5 x 11.1 ft)
PAYLOAD: UHF communications
payload; EHF communications
payload

UHF Follow-On was built for the US Navy's Space and Naval Warfare
Systems Command in San Diego, replacing the Fleet Satellite Communications (FLTSATCOM) and the Hughes-built Leasat spacecraft, its ultrahigh
frequency (UHF) satellite communications network. They support the Navy's
global communications network, serving ships at sea as well as other US
military fixed and mobile terminals. The first satellite, built by Hughes Space
and Communications Company (now Boeing), was launched in March 1993.

The UFO series is based on the three-axis-stabilized Hughes 601 bus.
The first seven satellites and F11 measure more than 18 m (60 ft) across the
two three-panel solar arrays. Each array on F8–F10 has four solar panels, so
the spacecraft is 22.9 m (75 ft) tip to tip. These arrays generate 2.5 kW of
power on F1–F3, 2.8 kW for F4–F7 and F11, and 3.0 kW for F8–F10. F8–F10
include a high-power, high-speed Global Broadcast Service (GBS) payload to
replaced the SHF payload. The first GBS payload was put into service in 1998.
It included four 130 W, 24 Mbps Ka-band transponders with three steerable
downlink spot-beam antennas, plus one steerable and one fixed uplink
antenna. This resulted in a 96 Mbps capability per satellite. F11 carries a new
ultrahigh frequency digital receiver with greater channel capacity than
previous UFOs. Three spacecraft give near-global coverage.

Wideband Global SATCOM (WGS, Wideband Gapfiller Satellite) USA

Communications

SPECIFICATION (WGS-1):

MANUFACTURER: Boeing Space and Intelligence Systems
LAUNCH DATE: October 10, 2007
ORBIT: GEO
LAUNCH SITE: Cape Canaveral, Florida
LAUNCHER: Atlas V
LAUNCH MASS: Approx. 5909 kg (13,000 lb)
BODY DIMENSIONS: 7.3 x 3.8 x 3.4 m (23.9 x 12.5 x 11.2 ft)
PAYLOAD: 8 X-band beams; 10 Ka-band beams

The WGS satellites of USAF Space Command are the key elements in a new high-capacity communications system for the US armed forces. The 13 kW WGS satellites are based on Boeing's 702 satellite bus. They will take over X-band communications provided by DSCS and the one-way Ka-band service provided by the Global Broadcast Service on UHF Follow-On satellites. WGS will also provide a new two-way Ka-band service.

Each WGS can route 2.1 to 3.6 Gbps of data, providing more than 10 times the communications capacity of the DSCS 3 satellite. The WGS payload can filter and route 4.875 GHz of instantaneous bandwidth. It has eight steerable and shapeable X-band beams formed by separate transmit and receive phased arrays, and 10 Ka-band beams served by independently steerable, diplexed antennas. Using reconfigurable antennas and a digital channelizer, WGS offers added flexibility of area coverage and is capable of connecting X-band and Ka-band users anywhere within its field of view.

Three Block I satellites will operate over the Pacific, Indian, and Atlantic oceans. They will be launched at six-monthly intervals. Two Block II satellites have been contracted. These will have a radio frequency bypass capability for intelligence and surveillance platforms requiring ultrahigh bandwidth and data rates—for example, unmanned aerial vehicles. Design life is 14 years.

Worldview USA

Military and commercial imagery

SPECIFICATION:

MANUFACTURER: Ball Aerospace
LAUNCH DATE: September 18, 2007
ORBIT: 493 km (306 mi), 97.5° inclination (Sun-synchronous)
LAUNCH SITE: Vandenberg, California
LAUNCHER: Delta II 7920
LAUNCH MASS: 2500 kg (5500 lb)
BODY DIMENSIONS: 3.6 x 2.5 m (11.8 x 8.2 ft)
PAYLOAD: WorldView-60 camera (WV60)

Worldview 1 was developed by DigitalGlobe, ITT, and Ball Aerospace with the help of $500 million of funding from the US Defense Department's National Geospatial-Intelligence Agency (NGA). Under this arrangement, the NGA is the main customer, but DigitalGlobe can sell images commercially as long as their resolution is no sharper than 0.5 m (1.6 ft). The spacecraft is the first element in the NextView reconnaissance satellite program, which combines commercial and defense imagery with the help of DOD funding.

The three-axis-stabilized spacecraft is based on the Ball Commercial Platform BCP-5000. It has twin solar arrays that provide 3.2 kW of power. Onboard data storage is 2199 Gb with X-band data downlink at 800 Mbps. Highest resolution for commercial use is 45 cm (17.7 in) panchromatic at nadir, with a 16 km (10 mi) wide imaging swath, and 51 cm (20.1 in) at 20° off-nadir. Imagery must be resampled to 50 cm (19.7 in) for non-US government customers.

Equatorial crossing is at 10.30 am local time on the descending node, with an average revisit time of 1.7 days. Rapid image-taking offers stereo capability. Images are used for precise map creation, detection of surface changes, and in-depth image analysis. Design life is at least seven years.

Xtar-Eur USA/SPAIN

Commercial and military communications

Xtar-Eur is the world's first satellite developed exclusively for commercial X-band services. The joint venture between Loral Space & Communications and Spanish company HISDESAT will provide steerable X-Band capacity for US, Spanish, and allied military and government applications.

Xtar-Eur is three-axis controlled and based on SS/L's 1300 platform. Power output is 5773 W. Station-keeping and orbital stability are maintained with bipropellant propulsion and momentum-bias systems. It carries 12 wideband, high-power X-band transponders. Onboard switching and multiple steerable beams allow users access to X-band capacity anywhere within its footprint.

The satellite provides X-band coverage from eastern Brazil and the Atlantic Ocean, across Europe, Africa, and the Middle East, as far east as Singapore. It offers critical X-band services when conventional military X-band systems are frequently at full capacity. Service began in April 2005. The Spanish Ministry of Defense was Xtar's first customer, leasing 238 MHz of X-band capacity until its primary satellite, Spainsat, entered service, after which time Xtar provided back-up capacity.

SPECIFICATION:

MANUFACTURER: Space Systems/Loral
LAUNCH DATE: February 12, 2005
ORBIT: 29° E (GEO)
LAUNCH SITE: Kourou, French Guiana
LAUNCHER: Ariane 5 ECA
LAUNCH MASS: 3630 kg (7986 lb)
BODY DIMENSIONS: 5.4 x 2.8 x 2.2 m (17.7 x 9.2 x 7.2 ft)
PAYLOAD: 12 X-band transponders

Scientific
Satellites:
Astronomy

AGILE (Astrorivelatore Gamma ad Immagini Leggero)

ITALY

Gamma-ray observatory

SPECIFICATION:

MANUFACTURER: Carlo Gavazzi Space/OHB-System
LAUNCH DATE: April 23, 2007
ORBIT: 550 km (341.8 mi), 2.5° inclination
LAUNCH MASS: 355 kg (780 lb)
LAUNCH SITE: Sriharikota, India
LAUNCHER: PSLV
DIMENSIONS: 1.7 x 1.7 x 2.6 m (5.5 x 5.5 x 8.6 ft) with solar panel
PAYLOAD: Gamma-ray Imaging Detector (GRID); Hard X-ray Imaging Detector (Super-AGILE); Mini-Calorimeter (MC)

AGILE is the first satellite in an Italian Space Agency (ASI) program of small scientific satellites. The mission is devoted to gamma-ray astrophysics, with the scientific and programmatic coparticipation of the Italian Institute of Astrophysics (INAF) and the Italian Institute of Nuclear Physics (INFN).

The hexagonal satellite has a fixed gallium arsenide solar panel attached to one side. The necessity of constant exposure to the Sun imposes some constraints on the pointing strategy and sky survey. A three-axis stabilization system allows satellite pointing to 0.5–1° and star sensors allow pointing to an accuracy of about 1 arcmin. A GPS transceiver ensures onboard timing accuracy within 1–2 microsec.

The onboard scientific instruments have optimal imaging capabilities in both the gamma-ray and hard X-ray energy ranges. The lightweight instruments are based on solid-state silicon detectors developed by INFN. It can detect and monitor gamma-ray sources in a region covering more than 20% of the sky. Over its two-year mission, AGILE will perform a complete sky survey of gamma-ray and hard X-ray sources.

Akari (Astro-F, IRIS—Infrared Imaging Surveyor) JAPAN

Infrared astronomy observatory

Akari (Light) was Japan's first infrared-ray astronomical satellite to perform an all-sky survey at
infrared wavelengths. Its instruments have much
greater sensitivity and higher resolution than
those achieved by the Infrared Astronomy Satellite
(IRAS). It has a 68.5 cm (27 in) telescope cooled to
6K (-267ºC), and observes in the wavelength range
from 1.7 (near-infrared) to 180 (far-infrared) μ. The
lightweight, silicon carbide primary mirror is
gold-coated. Akari's lifetime is based on the supply
of liquid helium stored in a cryosat to cool the
telescope and science instruments. The 170 l
(37 gal) of liquid helium ran out on August 26,
2007, ending far-infrared and mid-infrared
observations.

The all-sky survey observations began in May
2006, and the first survey was completed in
November 2006, covering about 80% of the sky.
The second mission phase was dedicated to
pointed observations, as well as gap-filling
observations for the all-sky survey. Akari is a JAXA
mission carried out with numerous partners.

SPECIFICATION:

MANUFACTURER: NEC Toshiba
LAUNCH DATE: February 21, 2006
ORBIT: 695 x 710 km (431.9 x
441.2 mi), 98.2º inclination
(Sun-synchronous)
LAUNCH MASS: 952 kg (2094 lb)
LAUNCH SITE: Uchinoura, Japan
LAUNCHER: M-V-8
BODY DIMENSIONS: 3.7 x 5.5 m
(12.1 x 18 ft) with solar panel
deployed
PAYLOAD: Far-Infrared Surveyor
(FIS); Infrared Camera (IRC)

Chandra X-Ray Observatory USA

X-ray observatory

Chandra (formerly known as the Advanced X-ray Astrophysics Facility) is the third in NASA's series of "Great Observatories," named in honor of the Nobel Prize-winning astrophysicist, Subrahmanyan Chandrasekhar. It was designed to study the composition of galaxies, stellar objects and interstellar phenomena as well as basic issues in theoretical physics. Its highly elliptical orbit has a period of 64 h, carrying it far outside Earth's radiation belts, allowing 55 h of uninterrupted observations during each orbit.

Chandra has eight times greater resolution and is able to detect sources more than 20 times fainter than any previous X-ray telescope. The telescope has four pairs of nearly cylindrical "nested" mirrors, ranging from 1.4 m to 0.68 m (4.6 to 2.2 ft) in diameter.

The spacecraft is three-axis stabilized, with six reaction wheels and four gyros. Two solar arrays generate over 2 kW of power. Two 1.8 Gb solid-state recorders can store 18.8 h of data each. Some loss of sensitivity in the CCDs on the ACIS instrument caused early concern. However, NASA decided in 2001 to extend the mission from 5 years to 10.

SPECIFICATION:

MANUFACTURER: TRW Space and Electronics Group
LAUNCH DATE: July 23, 1999
ORBIT: 9977 km x 138,400 km (6170 x 86,000 ml), 28.5° inclination
LAUNCH MASS: 4800 kg (10,560 lb)
LAUNCH SITE: Kennedy Space Center, Florida
LAUNCHER: Space Shuttle *Columbia* (STS-93)/IUS
DIMENSIONS: Solar arrays deployed, 13.8 x 19.5 m (45.3 x 64 ft)
PAYLOAD: Advanced Charged Couple Imaging Spectrometer (ACIS); High-Resolution Camera (HRC); High-Energy Transmission Grating (HETG); Low-Energy Transmission Grating (LETG)

COROT (COnvection, ROtation, and planetary Transits) FRANCE/EUROPE

Asteroseismology/exoplanet studies

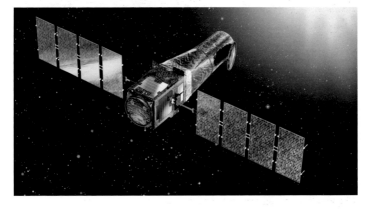

COROT is an astronomical mission to study the internal structure of stars and seek out planets around nearby stars. It will spend nearly three years studying the slight variations in the light output of about 120,000 stars. The variations caused by internal stellar vibrations will provide data on their internal structure, age, and composition—known as stellar seismology.

COROT is a CNES-led mission with ESA participation; the other international partners are Austria, Belgium, Brazil, Germany, and Spain. The three-axis-stabilized spacecraft is derived from the French Proteus bus and consists of a 30 cm (1 ft) telescope and a 4-CCD camera sensitive to tiny variations in the light intensity of stars.

Science observations began on February 3, 2007. The first observational phase, facing away from the center of our galaxy, lasted until April 2. Sun illumination conditions meant the spacecraft had to be rotated 180° to start observing the center of the Milky Way. On May 3, it was reported that COROT had obtained the first image of a giant planet orbiting a star in another solar system.

SPECIFICATION:

MANUFACTURER: Thales Alenia
LAUNCH DATE: December 27, 2006
ORBIT: 895 x 906 km (556 x 563 mi), 90° inclination
LAUNCH MASS: 630 kg (1386 lb)
LAUNCH SITE: Baikonur, Kazakhstan
LAUNCHER: Soyuz 2-1b
BODY DIMENSIONS: 4.1 x 2 m (13.5 x 6.5 ft)
PAYLOAD: Telescope; CCD camera

FUSE (Far Ultraviolet Spectroscopic Explorer) USA

Far-ultraviolet space observatory

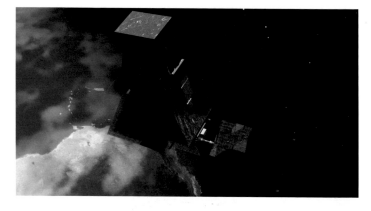

FUSE is a NASA Origins mission developed and operated by Johns Hopkins University in collaboration with NASA Goddard Space Flight Center, CNES, the Canadian Space Agency, the University of Colorado, and the University of California, Berkeley. It is the first large-scale space mission to be fully planned and operated by an academic department of a university. Its three-year prime mission was to obtain high-resolution spectra of faint galactic and extra-galactic objects at far-ultraviolet wavelengths. The box-shaped bus carries two solar arrays. Most of the spacecraft is taken up by a single instrument, mounted on top of the spacecraft, which contains four aligned telescopes. Light is focused by four curved mirrors onto four spectrograph gratings. In December 2001, two reaction wheels failed in quick succession, leaving the satellite in safe mode. Full science operations resumed in March 2002 after modification of flight-control software. A third reaction wheel was lost in 2004, but full operations resumed in January 2006. It also survived an anomaly with the only surviving reaction wheel in May 2007. The mission ended on October 18, 2007.

SPECIFICATION:

MANUFACTURER: Johns Hopkins University Applied Physics Laboratory
LAUNCH DATE: June 24, 1999
ORBIT: 745 x 760 km (463 x 472 mi), 25° inclination
LAUNCH MASS: 1360 kg (2992 lb)
LAUNCH SITE: Cape Canaveral, Florida
LAUNCHER: Delta II 7320
BODY DIMENSIONS: 0.9 x 0.9 x 1.3 m (3 x 3 x 4.3 ft)
PAYLOAD: FUSE spectrograph

GALEX (Galaxy Evolution Explorer) USA

Ultraviolet space observatory

SPECIFICATION:

MANUFACTURER: Orbital Sciences
Corporation
LAUNCH DATE: April 28, 2003
ORBIT: 694 x 700 km (431.2 x
435 mi), 28.99° inclination
LAUNCH MASS: 280 kg (617 lb)
LAUNCH SITE: Cape Canaveral,
Florida (air launch)
LAUNCHER: Pegasus XL
BODY DIMENSIONS: 1 x 2.5 m
(3 x 6.4 ft)
PAYLOAD: Telescope with 2
ultraviolet detectors

GALEX is a NASA Small Explorer-class mission. The orbiting space telescope is designed to observe a million distant galaxies in ultraviolet light in order to understand how the galaxies evolve. Led by the California Institute of Technology, GALEX is conducting several unique sky surveys, including an extragalactic, ultraviolet all-sky survey. During its 29-month prime mission, it produced the first comprehensive map of galaxies. By 2007, it had sent back thousands of ultraviolet images.

The spacecraft is a cylinder composed primarily of aluminium. The three-axis-stabilized spacecraft carries two gyroscope systems, four reaction wheels, and magnetic torquer bars and coil. Pointing is determined using a Sun sensor and a star tracker. Two fixed gallium arsenide solar arrays generate 290 W of power.

All science observations are made when the satellite is within Earth's shadow; the spacecraft recharges its batteries and communicates with ground stations during the daylight passes. Mission operations are conducted from a center in Dulles, Virginia, managed by Orbital Sciences Corporation.

GLAST (Gamma-Ray Large Area Space Telescope) USA

Gamma-ray space observatory

SPECIFICATION:

MANUFACTURER: General Dynamics C4 Systems (formerly Spectrum Astro)
LAUNCH DATE: January 31, 2008?
ORBIT: 565 km (350 mi), 28.5° inclination
LAUNCH MASS: 4277 kg (9429 lb)
LAUNCH SITE: Cape Canaveral, Florida
LAUNCHER: Delta 2920H-10
BODY DIMENSIONS: 2.8 x 2.5 m (9.2 x 8.2 ft)
PAYLOAD: Large-Area Telescope (LAT); GLAST Burst Monitor (GBM)

GLAST is a NASA astronomical satellite that will observe the universe in gamma rays. It has been developed in collaboration with the US Department of Energy and institutions in France, Germany, Japan, Italy, and Sweden.

GLAST will observe in the energy range 10 keV–300 GeV. The main instrument, the LAT, operates like a particle detector rather than a conventional telescope. With its very large field of view, the LAT sees about 20% of the sky at any time. In sky-survey mode, the LAT will cover the entire sky every three hours. It is 30 times more sensitive than any telescope in previous missions, enabling it to detect thousands of new gamma-ray sources. The LAT is expected to detect thousands of extremely bright quasars (blazers), accurately measuring their locations and the energy of their gamma rays.

The other instrument, the GBM, will detect roughly 200 gamma-ray bursts per year. The GBM will have a field of view several times larger than the LAT and will provide spectral coverage of gamma-ray bursts from the lower limit of the LAT down to 8 keV. The prime mission is five years, with a goal of 10 years of operations.

Gravity Probe B USA

Test of Einstein's general theory of relativity

SPECIFICATION:

MANUFACTURER: Stanford University
LAUNCH DATE: April 20, 2004
ORBIT: 640 x 646 km (390 x 401 mi), 90° inclination
LAUNCH MASS: 3145 kg (6933 lb)
LAUNCH SITE: Vandenberg, California
LAUNCHER: Delta II 7920
BODY DIMENSIONS: 6.4 x 2.6 m (21 x 8.7 ft)
PAYLOAD: GP-B telescope

Gravity Probe B was a NASA/Stanford University experiment to test two predictions of Einstein's general theory of relativity—warping of space and time due to the presence of the Earth, and "frame dragging" of space-time due to the Earth's rotation by measuring tiny changes in the direction of spin of four gyroscopes.

The spacecraft is built around a dewar filled with liquid helium. Along the dewar's central axis is the probe, a canister that contains heat-absorbing windows, a cryopump, and the science instrument assembly (SIA).

When each niobium-coated gyroscope rotates, a small magnetic field surrounds it. By monitoring the axis of this magnetic field, the direction the gyroscope's spin axis is pointing can be determined. The gyroscopes were so free from disturbance that they provided an almost perfect space-time reference system. Science operations began on August 28, 2004, and continued for 50 weeks. The spacecraft transmitted more than a terabyte of experimental data. Data collection ended on September 29, 2005, when the liquid helium was exhausted. First results were issued April 2007.

Herschel EUROPE

Infrared observatory

Herschel (formerly called Far-Infrared and Sub-millimeter Telescope, or FIRST) is an ESA infrared mission. Its 3.5 m (11.5 ft) mirror is the largest ever built for a space telescope. It will collect long-wavelength radiation from some of the coldest and most distant objects in the universe.

The service module contains the avionics, power distribution, communication, attitude control, and propulsion subsystems. On top is the payload module containing the liquid helium cryostat, science instruments, and telescope.

Herschel will launch on an Ariane 5 together with ESA's Planck spacecraft. The two spacecraft will separate after launch and be directly injected towards the second Lagrange point of the Sun–Earth system. Between 4 and 6 months after launch, Herschel will be injected into a large Lissajous orbit around the L2 point at a distance of around 1.5 million km (932,000 mi) from Earth. No insertion maneuver is needed. The orbit will be adjusted roughly once a month. Commissioning will take about six months. The nominal operational mission will last three years.

SPECIFICATION:

MANUFACTURER: Alcatel Alenia
LAUNCH DATE: July 2008?
ORBIT: L2 Lagrange point, Lissajous
LAUNCH MASS: 3300 kg (7260 lb)
LAUNCH SITE: Kourou, French Guiana
LAUNCHER: Ariane 5 ECA
BODY DIMENSIONS: 7.5 x 4 x 4 m (24.6 x 13.1 x 13.1 ft)
PAYLOAD: Heterodyne Instrument for the Far Infrared (HIFI); Photodetector Array Camera (PAC); Spectral and Photometric Imaging Receiver (SPIRE)

HETE-2 (High-Energy Transient Explorer-2) USA

Gamma-ray burst space observatory

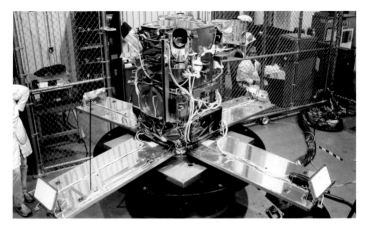

HETE-2 is the result of an international collaboration between the US, France, and Japan. It replaced the original HETE spacecraft that was lost in a launch failure in November 1996. The first mission devoted to gamma-ray bursts, HETE-2 was launched from the Kwajalein Missile Range in the Marshall Islands by a Pegasus dropped from an L-1011 aircraft.

HETE-2 has three main instruments sensitive to gamma and X-radiation; these share a common field of view. Within seconds of a burst, HETE-2 can calculate the precise location of a radiation burst. On the ground, a dedicated network of 12 listen-only burst-alert stations relays the data to the MIT control center. From there, information is transmitted to the Gamma Ray Burst Coordinate Distribution Network at NASA Goddard Space Flight Center, which can send the information to other observatories worldwide in 10–20 sec. HETE-2 allows astronomers to see a burst while it is still occurring and to study its development at various wavelengths. By March 2007, the reduced efficiency of the NiCd batteries was limiting operations.

SPECIFICATION:

MANUFACTURER: Massachusetts Institute of Technology (MIT)
LAUNCH DATE: October 9, 2000
ORBIT: 598 x 641 km (372 x 399 mi), 1.95° inclination
LAUNCH MASS: 124 kg (273 lb)
LAUNCH SITE: Kwajalein, Marshall Islands
LAUNCHER: Pegasus
BODY DIMENSIONS: 0.9 m x 0.7 m (2.9 x 2.2 ft)
PAYLOAD: French Gamma Telescope (FREGATE); wide-field X-ray monitor; soft X-ray camera; Radiation Belt Monitor (RBM); Global Positioning System (GPS) receiver

Hubble Space Telescope (HST) USA/EUROPE

Visible-infrared-ultraviolet space observatory

HST is a joint NASA–ESA project. It was designed for a 15-year lifetime, with periodic maintenance visits from Shuttle astronauts. The 2.4 m (7.5 ft) primary mirror collects light from distant objects, which is sent to various instruments. Fine guidance sensors lock onto guide stars to ensure extremely high pointing accuracy.

Poor images obtained after launch were due to a slightly misshapen mirror. This was corrected during Servicing Mission 1 (SM1) in December 1993. Since then, there have been three more servicing missions (SM2, February 1997; SM3A, December 1999; SM3B, March 2002) to replace instruments, gyros, solar panels, etc.

HST has been operating on two of its four gyroscopes since August 2005. The STIS failed in 2004. Most of the ACS failed in January 2007. A fifth and final SM (4) is planned for September 2008, when astronauts will install two new instruments and repair STIS. They will also attach a docking ring to the aft end of the telescope for a future deorbit motor. In 16 years, HST has taken roughly 750,000 exposures and studied about 24,000 celestial objects.

SPECIFICATION:

MANUFACTURER: Lockheed
LAUNCH DATE: April 24, 1990
ORBIT: 613 x 620 km (381 x 385 mi), 28.5° inclination
LAUNCH MASS: 10,843 kg (23,855 lb)
LAUNCH SITE: Kennedy Space Center, Florida
LAUNCHER: Space Shuttle *Discovery* (STS-31)
BODY DIMENSIONS: 13.3 x 4.7 m (43.6 x 15.4 ft)
PAYLOAD: Cosmic Origins Spectrograph (COS), 2008; Wide-field Camera 3 (WFC3), 2008; Advanced Camera for Surveys (ACS), 2002; Near Infrared Camera/Multi-Object Spectrometer (NICMOS), 1997; Space Telescope Imaging Spectrograph (STIS), 1997; Wide-field Planetary Camera 2 (WFPC2), 1993–2008; Corrective Optics Space Telescope Axial Replacement (COSTAR), 1993–2008; Wide-field/Planetary Camera (WF/PC), 1990–1993; Goddard High-Resolution Spectrograph (GHRS), 1990–1997; Faint Object Spectrograph (FOS), 1990–1997; Faint Object Camera (FOC), 1990–2002; High-Speed Photometer (HSP), 1990– 1993; Fine Guidance Sensors (FGS)

Infrared Space Observatory (ISO) EUROPE

Infrared space observatory

SPECIFICATION:

MANUFACTURER: Aerospatiale
LAUNCH DATE: November 17, 1995
ORBIT: 1038 x 70,578 km (645 x 43,856 mi), 5.2° inclination
LAUNCH MASS: 2498 kg (5496 lb)
LAUNCH SITE: Kourou, French Guiana
LAUNCHER: Ariane 44P
BODY DIMENSIONS: 5.3 x 3.6 x 2.8 m (17.4 x 11.8 x 9.2 ft)
PAYLOAD: ISO Camera (ISOCAM); ISO Photometer (ISOPHOT); Short Wavelength Spectrometer (SWS); Long Wavelength Spectrometer (LWS)

ISO was an ESA infrared observatory with instruments built by teams led from the UK, Germany, France, and the Netherlands. Operating at wavelengths from 2.5 to 240 μ, it was the most sensitive infrared satellite ever launched and made studies of the cool and dusty regions of the universe as well as of planets and comets.

The satellite comprised a large liquid-helium cryostat to cool the telescope and instruments to 1.8 K (-271°C), two fixed solar arrays that also acted as thermal shields, a 0.6 m (2 ft) telescope, and four scientific instruments.

After launch, ISO's hydrazine thrusters raised the perigee from 518 km (322 mi) to 1038 km (645 mi). Inside the Van Allen belts, ISO's detectors were unusable; observations were possible almost 17 h a day outside them. The planned lifetime was 20 months, but the helium coolant lasted until April 8, 1998. A few of the detectors in the SWS were used for 150 h after this, ending on May 10. ISO's final orbit was changed so that it will burn up in the atmosphere after 20–30 years. By May 1998, ISO had made nearly 30,000 scientific observations.

Integral (International Gamma-Ray Astrophysics Laboratory) EUROPE

Gamma-ray space observatory

Integral is a medium-class ESA science mission. It is the first space observatory to simultaneously observe celestial objects in gamma-rays, X-rays, and visible light. Its primary goals include mapping the remains of exploded stars and detecting matter-antimatter collisions.

Launched into a highly eccentric orbit by Proton K, Integral's thrusters were used to reach its operational 72-hour orbit on November 1, 2002. Orbit perigee will evolve over five years to 12,500 km (7767 mi) and an inclination of 87°.

Integral carries two main instruments (IBIS and SPI) for gamma-ray astrophysics (20 keV–10 MeV), plus two monitors (JEM-X and OMC) for X-ray and optical counterparts. IBIS is equipped with the first uncooled semiconductor gamma-ray camera. All instruments observe the same region of the sky simultaneously. Integral was the first observatory to show that most of the diffuse glow of gamma rays in the center of our galaxy is produced by about 100 individual sources. The mission has been extended until December 2008.

SPECIFICATION:

MANUFACTURER: Alenia Spazio
LAUNCH DATE: October 17, 2002
ORBIT: 9050 x 153,657 km (5624 x 95,480 ml), 52.25° inclination
LAUNCH MASS: 3958 kg (8708 lb)
LAUNCH SITE: Baikonur, Kazakhstan
LAUNCHER: Proton
BODY DIMENSIONS: 5 x 3.7 m (16.4 x 12.1 ft)
PAYLOAD: Imager on Board the Integral Satellite (IBIS); Spectrometer on Integral (SPI); Joint European X-ray Monitor (JEM-X); Optical Monitoring Camera (OMC)

James Webb Space Telescope (JWST) USA/EUROPE/CANADA

Infrared space observatory

Formerly known as the Next Generation Space Telescope, JWST is named after a former NASA administrator. It involves international cooperation between NASA, ESA, and the Canadian Space Agency (CSA), and its mission is to investigate the origin and evolution of galaxies, stars, and planetary systems. JWST is considered to be the successor of the Hubble Space Telescope. It is due to operate for at least five years.

After launch and transfer, JWST will operate at the L2 Lagrange point, 1.5 million km (932,000 mi) away on the night side of the Earth. It has a 6.5 m (21.3 ft) primary mirror made of silicon carbide. Light is fed to three instruments with superb imaging capability at visible and infrared wavelengths (0.6–28 μ). The observatory is always aligned so that the Sun, Earth, and Moon are on the bus side.

NASA is responsible for overall mission management and operations, including construction. It will also provide NIRCam via the University of Arizona. ESA will provide the launch and NIRSpec. MIRI is being built by a European consortium and NASA. FGS/TFI will be provided by the CSA.

SPECIFICATION:

MANUFACTURER: TRW (now Northrop Grumman)
LAUNCH DATE: 2013?
ORBIT: L2 Lagrange point, halo
LAUNCH MASS: 6200 kg (13,640 lb)
LAUNCH SITE: Kourou, French Guiana
LAUNCHER: Ariane 5 ECA
BODY DIMENSIONS: Sunshield, 22 x 12 m (72 x 39 ft)
PAYLOAD: Mid-Infrared Instrument (MIRI); Near-Infrared Camera (NIRCam); Near-Infrared spectrograph (NIRSpec); Fine Guidance Sensor/Tuneable Filter Imager (FGS/TFI)

Kepler USA

Search for exoplanets

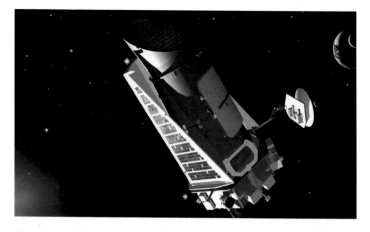

Kepler is a NASA Discovery-class mission designed to survey our region of the Milky Way galaxy to detect and characterize hundreds of Earth-size and smaller planets. The spacecraft is based on the design used for Deep Impact. Pointing at a single group of stars for the entire mission greatly increases the photometric stability and simplifies the spacecraft design. The spacecraft must be rolled 90° about the optical axis every three months to keep the solar array aligned with the Sun.

Kepler will operate from an Earth-trailing heliocentric orbit with a period of 372.5 days, meaning it slowly drifts away from the Earth and will be at a distance of up to 74.8 million km (46.5 million mi) after four years.

The single instrument, a photometer, has a 0.95 m (3.1 ft) aperture and a 1.4 m (4.6 ft) primary mirror. It has an array of 42 CCDs and a very large field of view. It will stare at the same star field for the entire mission, monitoring the brightness of more than 100,000 stars over four years, searching for planets in transit. Data will be stored onboard and transmitted to Earth about once a week.

SPECIFICATION:

MANUFACTURER: Ball Aerospace
LAUNCH DATE: February 2009?
ORBIT: Earth-trailing heliocentric
LAUNCH MASS: 1039 kg (2286 lb)
LAUNCH SITE: Cape Canaveral, Florida
LAUNCHER: Delta-II 2925-10L
BODY DIMENSIONS: 4.7 m (15.4 ft) from separation plane to top of sunshade; 2.7 m (8.9 ft) max. diameter of spacecraft bus
PAYLOAD: Photometer

LISA Pathfinder (Laser Interferometer Space Antenna)

EUROPE/USA

Gravitational wave technology demonstration

LISA Pathfinder is the name given to ESA's SMART-2 spacecraft. It will pave the way for the ESA–NASA LISA mission by flight-testing two competing concepts for gravitational wave detection.

The satellite comprises a science module and a propulsion module. The science spacecraft, with a mass of 420 kg (924 lb), contains the instruments and is covered with a single fixed solar array. Once in space, the test masses—two small cubes, each 5 cm (2 in) across and made of gold and platinum alloy—will float freely within the spacecraft. When subtle gravitational forces act on them, a laser beam will detect the way they change position to within a few thousandths of a billionth of a meter.

Launched into a parking orbit, LISA Pathfinder will use its propulsion system to progressively expand the orbit and reach the final operational Lissajous halo orbit around the L1 Lagrange point, 1.5 million km (932,000 mi) away in the direction of the Sun. After the last transfer burn is performed, the propulsion module will be jettisoned.

SPECIFICATION:

MANUFACTURER: EADS Astrium
LAUNCH DATE: 2010?
ORBIT: L1 Lagrange point, Lissajous orbit
LAUNCH MASS: 1900 kg (4180 lb)
LAUNCH SITE: Baikonur, Kazakhstan?
LAUNCHER: Rockot?
BODY DIMENSIONS: 2.1 x 1.0 m (6.9 x 3.3 ft)
PAYLOAD: LISA Technology Package (LTP); Disturbance Reduction System (DRS)

MOST (Microvariability and Oscillations of Stars)

CANADA

Study of stellar oscillations

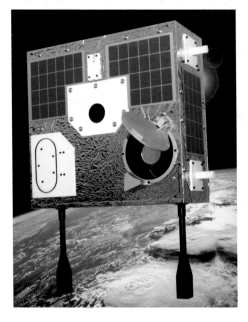

SPECIFICATION:

MANUFACTURER: Dynacon Enterprises/UTIAS Space Flight Lab
LAUNCH DATE: June 30, 2003
ORBIT: 818 x 832 km (508 x 517 mi), 98.7° inclination (Sun-synchronous)
LAUNCH MASS: 60 kg (132 lb)
LAUNCH SITE: Plesetsk, Russia
LAUNCHER: Rockot
BODY DIMENSIONS: 0.7 x 0.7 x 0.3 m (2.1 x 2.1 x 1 ft)
PAYLOAD: Photometer

MOST is Canada's first space telescope. The suitcase-sized microsatellite is designed to study tiny variations in light emissions from stars and to search for transiting extrasolar planets. Using a small telescope, it performs ultra-high-precision photometry of stars, down to the naked-eye limit of visibility (magnitude 6).

The cube-shaped satellite is powered by solar panels and oriented by a system of miniature reaction wheels and magneto-torquers. It carries one instrument, a photometer. The 15 cm (6 in) optical telescope feeds a CCD camera. One CCD is used for science measurements; the other is read out every second to track guide stars for satellite attitude control.

MOST was injected into a Sun-synchronous, near-polar orbit, with a period of about 100 min, that remains over the Earth's terminator. It has a continuous viewing zone spanning declinations from about -19 to +36°, in which a selected target star will remain observable for up to 60 days without interruption. Originally intended for a one-year mission, MOST continues to operate after four years.

Planck EUROPE

Cosmic microwave background radiation space observatory

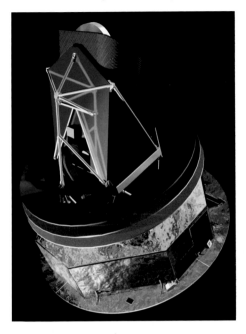

SPECIFICATION:

MANUFACTURER: Alcatel Space
LAUNCH DATE: July 2008?
ORBIT: L2 Lagrange point,
Lissajous orbit
LAUNCH MASS: 1800 kg (3960 lb)
LAUNCH SITE: Kourou, French
Guiana
LAUNCHER: Ariane 5 ECA
BODY DIMENSIONS: 4.2 x 4.2 m
(13.8 x 13.8 ft)
PAYLOAD: Low-frequency
Instrument (LFI); High-frequency
Instrument (HFI)

Planck is ESA's third medium-class science mission, originally named COBRAS/SAMBA. Its primary objective is to analyze, with the highest accuracy ever achieved, the cosmic microwave background radiation left over from the Big Bang. It will also obtain information on dust and gas in the Milky Way, as well as in other galaxies.

The satellite is spin-stabilized at 1 rpm. The payload consists of a 1.5 m (4.9 ft) telescope offset by 85° from the spin axis in order to scan the sky. The telescope focuses radiation onto two arrays of highly sensitive detectors. Below the payload module is the octagonal service module, containing the power, propulsion, and other satellite systems.

Planck will be launched together with Herschel. They will separate shortly after the launch and proceed independently to different orbits about the L2 Lagrange point, 1.5 million km (932,000 mi) away on the Earth's night side. After a journey of 4–6 months, Planck will make a major maneuver and enter a small Lissajous orbit around L2. Planck's operational lifetime will be about 21 months.

Rossi X-ray Timing Explorer (RXTE) USA

Study of variable X-ray sources

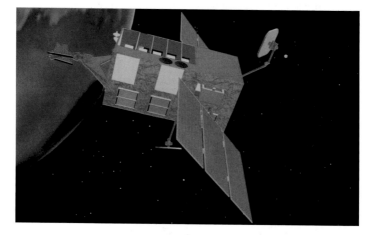

RXTE, named after the Italian-American astronomer Bruno Rossi, is a NASA Explorer-class mission to study the variability of X-ray sources with unprecedented time resolution, in combination with moderate spectral resolution.

It is designed to enable flexible operations through rapid pointing, high data rates, and nearly continuous receipt of data at the Science Operations Center NASA–Goddard.

RXTE carries two pointed instruments, the PCA and HEXTE, which are equipped with collimators yielding a 1° field of view. An ASM scans about 80% of the sky every orbit, allowing monitoring over 90 min or longer. RXTE is highly maneuverable, with a slew rate of more than 6°/min. The PCA/HEXTE can be pointed anywhere in the sky to an accuracy of less than 0.1°.

RXTE has made important observations of millisecond pulsars, black holes, and the galactic X-ray background. RXTE's expected orbital lifetime was 10 years, but it was still operational in July 2007.

SPECIFICATION:

MANUFACTURER: NASA-Goddard
LAUNCH DATE: December 30, 1995
ORBIT: 565 x 583 km (351 x 362 mi), 23° inclination
LAUNCH MASS: 3045 kg (6600 lb)
LAUNCH SITE: Cape Canaveral, Florida
LAUNCHER: Delta II 7920
BODY DIMENSIONS: 1.8 x 1.8 x 5.4 m (5.9 x 5.9 x 17.7 ft)
PAYLOAD: Proportional Counter Array (PCA); High-energy X-ray Timing Experiment (HEXTE); All-Sky Monitor (ASM)

Spitzer Space Telescope USA

Infrared space observatory

SPECIFICATION:

MANUFACTURER: Lockheed Martin
Space Systems
LAUNCH DATE: August 25, 2003
ORBIT: Earth-trailing heliocentric,
149 million x 152.4 million km
(92.6 million x 94.7 million mi),
1.1° inclination
LAUNCH MASS: 865 kg (1907 lb)
LAUNCH SITE: Cape Canaveral,
Florida
LAUNCHER: Delta II 7920H
BODY DIMENSIONS: 4.5 x 2.1 m
(14.6 x 6.9 ft)
PAYLOAD: Infrared Spectrograph
(IRS); Multiband Imaging
Photometer for Spitzer (MIPS);
Infrared-detector Array Camera
(IRAC)

Spitzer, formerly the Space Infrared Telescope Facility (SIRTF), was renamed
in honor of US physicist Lyman Spitzer. It is the fourth of NASA's Great
Observatories. Able to detect radiation at -180 μ, it is returning
unprecedented views of the cool objects in the universe, including dust
clouds related to the formation of stars and planets.

The tube-shaped cryogenic assembly includes a telescope with an 0.8 m
(2.8 ft) beryllium mirror and three scientific instruments.

Since the two solar arrays are fixed, Spitzer must be pointed within 120°
of the Sun. It observes a strip of sky roughly perpendicular to the Sun, with
about 35% of the sky visible at any one time. Spitzer was boosted from Earth
parking orbit into an Earth-trailing heliocentric orbit, which causes its
distance from the Earth to increase by about 15 million km (9.3 million mi)
per year. This allows uninterrupted viewing of a large portion of the sky and
provides a very good thermal environment. The first infrared images were
released on December 18, 2003.

Suzaku (Astro-E2) JAPAN/USA

X-ray space observatory

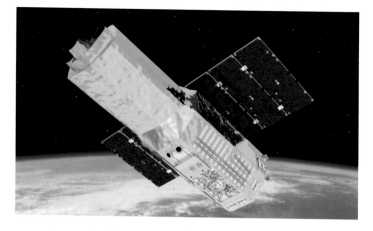

Astro-E2, the fifth Japanese X-ray astronomy satellite to observe hot and active regions in the universe, replacing the original Astro-E satellite, lost during launch in February 2000. It was named Suzaku after a legendary Japanese red bird.

NASA provided the core instrument, the first X-ray microcalorimeter array to be placed in orbit. It measured the heat created by individual X-ray photons and operated at a temperature only 0.1° above absolute zero. It was cooled by a three-stage system developed by NASA–Goddard and ISAS. On August 8, 2005, a major liquid helium leak led to the loss of the XRS. NASA also provided the five telescopes required to focus X-rays on the XRS and X-ray cameras. Japan provided the four lightweight X-ray cameras and a high-energy X-ray detector. The image resolution of the XIS is about 2 arcmin, but the cameras are extremely efficient at collecting X-ray photons. The HXD covers high-energy (hard) X-rays from 0.5 keV to 600keV at the highest sensitivity ever achieved. The XIS detector took its first images on August 13, 2005, and the HXD on August 19.

SPECIFICATION:

MANUFACTURER: JAXA-ISAS
LAUNCH DATE: July 10, 2005
ORBIT: operational, 562 x 573 km (349 x 356 mi), 31.4° inclination
LAUNCH MASS: 1680 kg (3700 lb)
LAUNCH SITE: Uchinoura, Japan
LAUNCHER: M-V-6
BODY DIMENSIONS: 5 x 2 m (16.4 x 6.6 ft)
PAYLOAD: X-ray Spectrometer (XRS); Hard X-ray Detector (HXD); 4 X-ray Imaging Spectrometers (XIS)

SWAS (Submillimeter Wave Astronomy Satellite) USA

Submillimeter space observatory

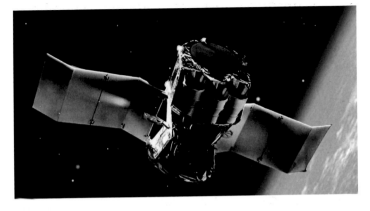

SWAS, the third of NASA's Small Explorer missions, is a three-axis-stabilized radio observatory with a pointing accuracy of 38 arcsec and jitter of less than 19 arcsec. It was designed to study star formation by determining the composition of interstellar clouds and examining how these clouds cool as they collapse to form stars and planets. It was the first satellite to detect submillimeter radiation emitted by water and oxygen molecules inside dense, hot vapor clouds between stars. The SWAS Science Operations Center is located at the Harvard-Smithsonian Center for Astrophysics.

Each December, SWAS returns to the orbital configuration that it had at launch; targets visible to the satellite at the beginning of the mission become available once more for observation. During its three-year nominal life span, it made detailed 1° x 1° maps of giant molecular clouds. It was turned off on July 21, 2004, but reactivated June–August 2005 to monitor water emissions from Comet Tempel 1 during the Deep Impact collision. SWAS is now in hibernation.

SPECIFICATION:

MANUFACTURER: Ball Aerospace
LAUNCH DATE: December 5, 1998
ORBIT: 637 x 653 km (396 x 406 ml), 70° inclination
LAUNCH MASS: 288 kg (637 lb)
LAUNCH SITE: Vandenberg, California (air launch)
LAUNCHER: Pegasus XL
BODY DIMENSIONS: 1 m (3.2 ft)
PAYLOAD: Acousto-optical spectrometer; Submillimeter Wave Receiver

Swift USA

Gamma-ray burst space observatory

SPECIFICATION:

MANUFACTURER: Spectrum Astro
LAUNCH DATE: November 20, 2004
ORBIT: 584 x 604 km (362 x 375 mi), 20.6 deg inclination
LAUNCH MASS: 1331 kg (2934 lb)
LAUNCH SITE: Cape Canaveral, Florida
LAUNCHER: Delta II 7320-10
BODY DIMENSIONS: 5.1 x 1.7 m (16.6 x 5.6 ft)
PAYLOAD: Burst Alert Telescope (BAT); X-ray Telescope (XRT); UV/Optical Telescope (UVOT)

Swift is a NASA Medium Explorer (MIDEX) mission. The space observatory is dedicated to the study of gamma-ray bursts (GRBs). Its main mission objectives are to determine the origin of gamma-ray bursts; study how the blast wave evolves; and perform the first sensitive, hard X-ray survey of the sky.

The three instruments work together to observe GRBs and afterglows in the gamma-ray, X-ray, ultraviolet, and optical wavebands. The BAT includes the largest "coded aperture mask" ever built and detects and acquires high-precision locations for GRBs, relaying an estimate position to the ground within 15 sec. 20–75 seconds after the burst detection, the spacecraft realigns itself to bring the GRB within the narrow field of view of the X-ray and UV/optical telescopes.

The Mission Operations Center is located at Pennsylvania State University. It receives almost all data via a transmission station in Kenya maintained by the Italian Space Agency. The first GRB detection was made on January 17, 2005. During its nominal two-year mission, Swift observed over 200 bursts, including the most distant GRB, 13 billion light years away.

WMAP (Wilkinson Map Anisotropy Probe) USA

Cosmic microwave background radiation space observatory

SPECIFICATION:

MANUFACTURER: NASA–Goddard
LAUNCH DATE: June 30, 2001
ORBIT: L2 Lagrange point,
Lissajous orbit
LAUNCH MASS: 840 kg (1848 lb)
LAUNCH SITE: Cape Canaveral,
Florida
LAUNCHER: Delta II 7425-10
BODY DIMENSIONS: 3.8 x 5 m
(12.5 x 16.4 ft)
PAYLOAD: 2 differential
microwave radiometers

WMAP is NASA's second Medium Explorer (MIDEX) mission. It was named in honor of US cosmologist David T. Wilkinson. Its primary mission was to observe in great detail the fluctuations in the cosmic microwave background (CMB) radiation left over from the Big Bang.

The satellite's main feature is a pair of 1.4 x 1.6 m (4.6 x 5.2 ft) back-to-back telescopes that collect microwave radiation from two spots on the sky roughly 140° apart and feed it to 10 receivers directly beneath the optics.

WMAP is the first mission to operate from an L2 orbit. It was placed into a highly elliptical parking orbit, then used its thrusters to maneuver for a lunar gravity assist. The lunar swing-by occurred on July 30, and on October 1 WMAP reached its Lissajous orbit about the L2 Lagrange point, 1.5 million km (932,000 mi) from Earth. WMAP scans the full sky every six months.

The original observation period was 24 months, but the mission has been extended to six years. WMAP has completed four full sky scans and is now primarily focused on the much weaker polarized signals from the CMB.

WISE (Wide-field Infrared Survey Explorer) USA

Infrared space observatory

SPECIFICATION:

MANUFACTURER: Ball Aerospace
LAUNCH DATE: June 2009?
ORBIT: 500 km (310 mi),
97.3° inclination (Sun-
synchronous)
LAUNCH MASS: 750 kg (1650 lb)
LAUNCH SITE: Vandenberg,
California
LAUNCHER: Delta II 7320-10
BODY DIMENSIONS: Not known
PAYLOAD: 4 focal plane arrays

WISE is a NASA Medium Explorer (MIDEX) mission designed to scan the entire sky in infrared light. It will detect cool stars close to the Sun and main belt asteroids, study star formation, and help to identify objects worth observing with the James Webb Space Telescope. The spacecraft derives from the Ball Aerospace NextSat built for the Orbital Express mission. The spacecraft will be three-axis stabilized, with body-fixed solar arrays, and will use a high-gain antenna to transmit to ground through the TDRSS geostationary system.

The observatory has a 0.4 m (1.3 ft) telescope with a 47 arcmin field of view. The infrared instrument will survey the sky simultaneously in two near-infrared channels, 3.3 and 4.7 μ, and two mid-infrared channels, 12 and 23 μ.

WISE will always keep its solar panels to the Sun. The instrument takes images every 11 sec, while the spacecraft maintains a continuous pitch rate that matches the orbit pitch. In six months, the entire sky will be imaged, with eight or more exposures at each position. Planned lifetime is seven months, including six months of operations.

XMM-Newton (X-ray Multi-Mirror) EUROPE

X-ray space observatory

XMM is ESA's second Cornerstone mission. Its name is derived from the X-ray Multi-Mirror design and honors Sir Isaac Newton. It is the most sensitive X-ray observatory ever launched, using over 170 wafer-thin cylindrical mirrors to gather and focus X-rays. The total surface area of the nested mirrors is more than 120 m² (1290 ft²). Two of the three X-ray mirror modules are fitted with RGSs for the most detailed analysis of the X-ray energies.

Forty-six companies in 14 European countries, and one in the USA, contributed to XMM's construction. The lead scientific investigators come from the Netherlands and the UK. The three-axis-stabilized spacecraft has a pointing accuracy of 1 arcsec, four reaction wheels, two star trackers, four inertial measure-ment units, and three fine Sun sensors, and three Sun acquisition sensors provide attitude control.

XMM's orbit takes it almost one third of the way to the Moon, providing long, uninterrupted views of celestial objects. XMM spends most of its time south of the equator, traveling quite slowly out to distances of more than 100,000 km (62,140 mi), well clear of the Earth's radiation belts.

SPECIFICATION:

MANUFACTURER: Dornier Satellite Systems
LAUNCH DATE: December 10, 1999
ORBIT: 7365 x 113,774 km (4577 x 70,698 mi), 38.9° inclination
LAUNCH MASS: 3764 kg (8280 lb)
LAUNCH SITE: Kourou, French Guiana
LAUNCHER: Ariane 5
DIMENSIONS: On-orbit, 10.1 x 16.2 m (33 x 53 ft)
PAYLOAD: 3 European Photon Imaging Cameras (EPIC); 2 Reflection Grating Spectrometers (RGS); Optical Monitor (OM)

Scientific Satellites: Solar system

ACE (Advanced Composition Explorer) USA

Solar wind studies

ACE is in orbit about the L1 point, about 1.5 million km (932,000 mi) from the Earth in the direction of the Sun. The spacecraft has four solar arrays, two of which have magnetometer booms attached. The launch mass included 189 kg (416 lb) of hydrazine fuel for orbit insertion and maintenance. The spacecraft spins at 5 rpm, with the spin axis generally pointed along the Earth–Sun line.

ACE carries a magnetometer, six high-resolution sensors, and three monitoring instruments to sample low-energy particles of solar origin and high-energy galactic particles. ACE performs measurements over a wide range of energy and nuclear mass, under all solar wind flow conditions and during both large and small particle events including solar flares. When reporting space weather, ACE can provide about 1 h of advance warning of geomagnetic storms that are likely to affect Earth.

Now in orbit for 10 years, the satellite is still working very well, with the exception of the SEPICA instrument. The spacecraft has enough propellant on board to maintain an orbit at L1 until about 2019.

SPECIFICATION:

MANUFACTURER: Johns Hopkins University Applied Physics Laboratory
LAUNCH DATE: August 25, 1997
ORBIT: L1 libration point (halo)
LAUNCH MASS: 785 kg (1727 lb)
LAUNCH SITE: Cape Canaveral, Florida
LAUNCHER: Delta II 7920
BODY DIMENSIONS: 1.6 x 1 m (5.3 x 3.3 ft)
PAYLOAD: Cosmic Ray Isotope Spectrometer (CRIS); Solar Isotope Spectrometer (SIS); Ultra-Low-Energy Isotope Spectrometer (ULEIS); Solar Energetic Particle Ionic Charge Analyzer (SEPICA); Solar Wind Ion Mass Spectrometer (SWIMS); Solar Wind Ionic Composition Spectrometer (SWICS); Electron, Proton, and Alpha Monitor (EPAM); Solar Wind Electron, Proton, and Alpha Monitor (SWEPAM); Magnetometer (MAG); Real-time Solar Wind (RTSW)

BepiColombo EUROPE/JAPAN

Two Mercury orbiters

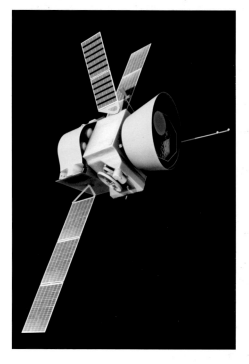

SPECIFICATION:

MANUFACTURER (MPO): EADS Astrium
MANUFACTURER (MMO): JAXA
LAUNCH DATE: August 2013?
MERCURY ORBITS (MPO): 400 x 1500 km (248 x 932 mi), polar
MERCURY ORBITS (MMO): 400 x 12,000 km (248 x 7456 mi), polar
LAUNCH MASS: 2300 kg (5060 lb)
LAUNCH SITE: Baikonur, Kazakhstan
LAUNCHER: Soyuz 2-1B/Fregat-M
BODY DIMENSIONS (MPO): 1.8 x 1.5 m (5.9 x 4.9 ft)
BODY DIMENSIONS (MMO): 1.8 x 0.9 m (5.9 x 2.9 ft)
PAYLOAD (MPO): BepiColombo Laser Altimeter; Italian Spring Accelerometer; Mercury Magnetometer; Mercury Thermal Infrared Spectrometer; Mercury Gamma-Ray and Neutron Spectrometer; Mercury Imaging X-ray Spectrometer; Probing of Hermean Exosphere by Ultraviolet Spectroscopy; Search for Exosphere Refilling and Emitted Neutral Abundances; Spectrometers and Imagers for MPO BepiColombo; Solar Intensity X-ray Spectrometer
PAYLOAD (MMO): Mercury Magnetometer; Mercury Plasma Particle Experiment; Plasma Wave Instrument; Mercury Sodium Atmospheric Spectral Imager; Mercury Dust Monitor

The BepiColombo mission, named after an Italian mathematician and engineer, comprises three modules that will be launched as a single spacecraft. The Mercury Planetary Orbiter (MPO) will be under ESA responsibility, while JAXA is responsible for the octagonal Mercury Magnetospheric Orbiter (MMO). ESA will also oversee the Mercury Transfer Module, which will provide the solar-electric and chemical propulsion required to reach Mercury.

The 500 kg (1100 lb) MPO will carry 11 scientific instruments. The 250 kg (550 lb) MMO will carry five advanced scientific experiments, one European and four from Japan, to investigate the planet's magnetic field. The spacecraft will take six years to reach Mercury, using gravity assists from the Moon, Earth, Venus, and Mercury. Observations will continue for at least one Earth year.

Various thermal protection measures will be used, including a multilayer insulating blanket, a radiator whose design makes it less sensitive to the thermal infrared radiation emitted by the planet's surface.

Cassini USA

Saturn orbiter

Cassini was the first spacecraft to orbit Saturn. It was named after the French/Italian astronomer Jean-Dominique Cassini. Launched together with ESA's Huygens probe, it was the largest US planetary spacecraft ever flown. To reach Saturn, Cassini-Huygens used a series of gravity-assist maneuvers, with two swingbys of Venus (April 27, 1998, and June 24, 1999), one of Earth (August 18, 1999), and one of Jupiter (December 30, 2000). The four-year primary mission included 75 orbits around Saturn, with 45 close flybys of Titan.

On July 1, 2004, Cassini-Huygens entered the Saturnian system. During orbit insertion, it passed through the gap between Saturn's F and G rings, using its high gain antenna as a dust shield. On December 25, 2004, towards the end of its third orbit around Saturn, the Cassini orbiter jettisoned the Huygens probe and relayed data back to Earth during Huygens's descent to the surface of Titan. Cassini carries 11 science instruments. The Italian Space Agency (ASI) contributed its high-gain antenna and other radio subsystem equipment.

SPECIFICATION:

MANUFACTURER: NASA-JPL
LAUNCH DATE: October 15, 1997
SATURN ORBIT (INITIAL): 80,731 x 9.1 million km (50,158 x 5.6 million mi), 11.5° inclination
LAUNCH MASS: 5257 kg (11,565 lb)
LAUNCH SITE: Cape Canaveral, Florida
LAUNCHER: Titan IVB-Centaur
BODY DIMENSIONS: 6.8 x 4 m (22.3 x 13.1 ft)
PAYLOAD: Composite Infrared Spectrometer (CIRS); Imaging Science Subsystem (ISS); Ultraviolet Imaging Spectrograph (UVIS); Visible and Infrared Mapping Spectrometer (VIMS); Cassini Plasma Spectrometer (CAPS); Cosmic Dust Analyzer (CDA); Ion and Neutral Mass Spectrometer (INMS); Magnetometer (MAG); Magnetospheric Imaging Instrument (MIMI); Radio and Plasma Wave Science (RPWS); Cassini Radar

Chandrayaan-1 INDIA

Lunar orbiter

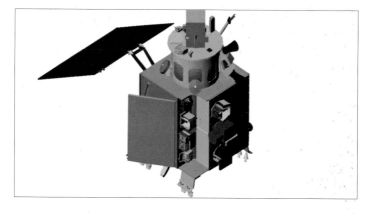

Chandrayaan-1, the first deep space mission of the Indian Space Research Organization (ISRO), is a lunar orbiter. Its goals include expanding scientific knowledge of the Moon, upgrading India's technological capability, and providing challenging opportunities for planetary research for the younger generation. It will provide high-resolution remote sensing and three-dimensional mapping of the Moon.

The spacecraft is based on technologies developed for the IRS and INSAT satellites. At lunar orbit the orbiter will weigh 590 kg (1298 lb), including a 55 kg (120 lb) package of scientific instruments and a 30 kg (65 lb) impactor designed to penetrate the lunar surface. The Moon Impact Probe carries a radar altimeter, a video imaging system, and a mass spectrometer for measuring the composition of the tenuous lunar atmosphere during descent.

Chandrayaan-1 will be launched by a modified PSLV rocket that will place the spacecraft into a 240 x 24,000 km (149 x 14,913 mi) Earth orbit. Later, the orbit will be raised to a Moon-rendez-vous orbit by multiple in-plane perigee maneuvers. The designed mission lifetime is two years.

SPECIFICATION:

MANUFACTURER: ISRO
LAUNCH DATE: April 2008
ORBIT: Lunar, 100 km (62 mi), polar
LAUNCH MASS: 1304 kg (2869 lb)
LAUNCH SITE: Sriharikota, India
LAUNCHER: PSLV-XL
BODY DIMENSIONS: 1.5 x 1.5 x 1.5 m (4.9 ft)
PAYLOAD: Moon Mineralogy Mapper (M3), Miniature Synthetic Aperture Radar (Mini-SAR); Moon Impact Probe (MIP); High-Energy X-ray Spectrometer (HEX); Terrain-Mapping Stereo Camera (TMC); Hyperspectral Imager (HySI); Lunar Laser Ranging Instrument (LLRI); Radiation Dose Monitor Experiment (RADOM); Sub-KeV Atom Reflecting Analyzer (SARA); Near-InfraRed Spectrometer (SIR-2); Chandrayaan-1 X-ray Spectrometer (C1XS)

Cluster II (Rumba, Salsa, Samba, Tango) EUROPE

Magnetosphere/Earth–Sun connection

SPECIFICATION:

MANUFACTURER: Dornier Satellite
Systems
LAUNCH DATES: July 16, 2000,
FM7 (Samba) and FM6 (Salsa);
August 9, 2000, FM5 (Rumba)
and FM8 (Tango)
ORBIT: 17,200 x 120,500 km
(10,688 x 74,877 mi), 90.7°
inclination
LAUNCH MASS: 1200 kg (2640 lb)
(x 4)
LAUNCH SITE: Baikonur,
Kazakhstan
LAUNCHER: Soyuz-Fregat (2)
BODY DIMENSIONS: 2.9 x 1.3 m
(9.5 x 4.3 ft)
PAYLOAD: Active Spacecraft
Potential Control (ASPOC);
Cluster Ion Spectrometry
experiment (CIS); Digital Wave
Processing experiment (DWP);
Electron Drift Instrument (EDI);
Electric Field and Wave experi-
ment (EFW); Fluxgate Magneto-
meter (FGM); Plasma Electron
and Current Experiment (PEACE);
Research with Adaptive Particle
Imaging Detectors (RAPID);
Spatio-Temporal Analysis of Field
Fluctuations (STAFF); Wide-Band
Data Instrument (WBD); Waves
of High Frequency and Sounder
for Probing of Electron Density
by Relaxation (WHISPER)

The original Cluster mission was lost during a launch failure of the Ariane 5 rocket on June 4, 1996. The four identical Cluster II spacecraft were rebuilt and launched in pairs one month apart. They were placed into a 240 x 18,000 km (149 x 11,185 mi) parking orbit and then maneuvered to their operational orbits.

Each cylindrical spacecraft is spin-stabilized, rotating at 15 rpm. Power is provided by six exterior solar panels. Propellant accounted for more than half the launch weight; most of this was consumed soon after launch and in complex maneuvers to reach operational orbits. The scientific payload—11 instruments to study space weather and the magnetosphere—weighs 71 kg (156 lb). Scientific operations started on February 1, 2001.

During the mission, spacecraft separation has ranged from 200 km to 10,000 km (124 to 6214 mi). By flying in a close, tetrahedral formation, the four spacecraft have provided the first small-scale, three-dimensional data on the physical processes views of near-Earth space. The five-year mission has been extended until December 2009.

Dawn USA

Asteroid orbiter

Dawn will be the first spacecraft to orbit two of the largest main belt asteroids, Ceres and Vesta. When Dawn began development in February 2004, the laser altimeter and magnetometer were not included, leaving three science instruments, one each provided by Germany, Italy, and Los Alamos in the US. On March 2, 2006, NASA canceled the project, but after an appeal the decision was reversed.

Dawn launched with 425 kg (937 lb) of xenon. Its ion propulsion subsystem electrically accelerates ionized xenon gas, emitting it at very high speed from any one of three ion thrusters. The ion engine is also used for attitude control. The spacecraft's main antenna is 1.5 m (5 ft) in diameter, and three smaller antennas are also available. Electrical power is provided by two solar arrays almost 8.3 m (27.2 ft) long. These generate over 10 kW at Earth's distance from the Sun.

Dawn will arrive at Vesta in August 2011, after a Mars gravity assist in February 2009. It will leave Vesta in May 2012, reaching Ceres, the largest asteroid, in February 2015. The primary mission will end in July 2015.

SPECIFICATION:

MANUFACTURER: Orbital Sciences Corporation
LAUNCH DATE: September 27, 2007
ORBIT: Heliocentric
LAUNCH MASS: 1218 kg (2680 lb)
LAUNCH SITE: Cape Canaveral, Florida
LAUNCHER: Delta II 7925-H
BODY DIMENSIONS: 1.6 x 1.3 x 1.8 m (5.4 x 4.2 x 5.8 ft)
PAYLOAD: Framing camera; visual and infrared spectrometer; gamma-ray and neutron spectrometer

Deep Impact USA

Comet flyby/impact

Deep Impact, the eighth mission in NASA's Discovery program, was designed to investigate Comet Tempel 1 by sending an impactor into its nucleus.

After a journey of 173 days and 431 million km (268 million mi), the flyby took place on July 4, 2005. One day out from the comet, the flyby spacecraft deployed the impactor. At the time of impact, the flyby spacecraft was 8606 km (5348 mi) from the comet. Its closest approach to the nucleus took place 14 min after the impact, at a distance of about 500 km (310 mi).

Both spacecraft were designed to perform autonomous onboard navigation. From 2 h before impact, their "autonav" software began taking images at 15 sec intervals. The impactor's thrusters fine-tuned its flight path so that impact occurred in a sunlit area visible by the flyby space-craft. The impactor's final image was transmitted 3 sec before impact, about 30 km (18.6 mi) above the surface. Approximately 4500 images were returned by the three cameras, but the impact crater was hidden by the huge cloud of debris.

SPECIFICATION:

MANUFACTURER: Ball Aerospace & Technologies Corp.
LAUNCH DATE: January 12, 2005
ORBIT: Heliocentric
LAUNCH MASS: 601 kg (1325 lb)
LAUNCH SITE: Cape Canaveral, Florida
LAUNCHER: Delta II 7925 with Star 48 upper stage
BODY DIMENSIONS (FLYBY CRAFT): 3.3 x 1.7 x 2.3 m (10.8 x 5.6 x 7.5 ft)
BODY DIMENSIONS (IMPACTOR): 1 x 1 m (3.3 ft)
PAYLOAD (FLYBY CRAFT): High-resolution instrument (multispectral camera and infrared spectrometer); medium-resolution instrument (camera)
PAYLOAD (IMPACTOR): Medium-resolution instrument (camera); impactor targeting sensor

Deep Space 1 USA

Technology demonstration/comet and asteroid flybys

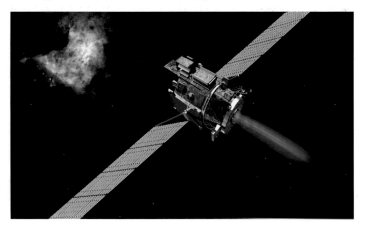

Deep Space 1 was the first spacecraft in NASA's New Millennium program. Its primary mission was to test 12 new advanced technologies. Most of these technologies were validated during the first few months of flight. It then passed 26 km (16 mi) from asteroid 9969 Braille on July 29, 1999. The mission was extended to include a 2170 km (1348 mi) flyby of Comet Borrelly on September 22, 2001.

The probe carried a xenon ion engine, an autonomous navigation system, and two gallium arsenide solar arrays that concentrated sunlight for extra power. Other new technologies included two more autonomy experiments, a small transponder, a Ka-band solid-state power amplifier, and experiments in low-power electronics, power switching, and multifunctional structures. Communications were via a high-gain antenna, three low-gain antennas, and a Ka-band antenna. All were mounted on top of the spacecraft except one low-gain antenna mounted on the bottom. The ion engines were switched off on December 19, 2001, having consumed 90% of the xenon. The radio receiver was left on, in case future generations want to contact the spacecraft.

SPECIFICATION:

MANUFACTURER: Spectrum-Astro
LAUNCH DATE: October 24, 1998
ORBIT: Heliocentric
LAUNCH MASS: 489 kg (1075 lb)
LAUNCH SITE: Cape Canaveral, Florida
LAUNCHER: Delta II 7326
BODY DIMENSIONS: 2.5 x 2.1 x 1.7 m (8.2 x 6.9 x 5.6 ft)
PAYLOAD: Miniature Integrated Camera-Spectrometer (MICAS); Plasma Experiment for Planetary Exploration (PEPE); Solar Concentrator Arrays with Refractive Linear Element Technology (SCARLET II); remote agent; Autonomous Navigation (AutoNav); Beacon Monitor Operations experiment; Ka-band solid-state power amplifier; small deep-space transponder

Double Star (DS-1, DS-2/Tan Ce (TC)/Explorer)

CHINA-EUROPE

Magnetosphere, Sun–Earth connection

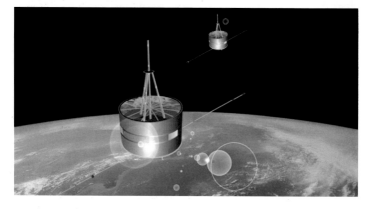

The China–ESA Double Star project used two spacecraft to study the Earth's magnetosphere, in concert with ESA's four Cluster spacecraft. Their orbits were much closer to Earth than Cluster, and were originally synchronized so that all six spacecraft could study simultaneously the same region of near-Earth space. In China they were known as Tan Ce (Explorer). DS-1 was in an elliptical, low inclination orbit. DS-2 was in a highly elliptical polar orbit with a 12 month prime mission. During 2007, the orbital planes of Cluster and DS-1 were separated by around 60° in azimuth or 'local time' to allow different types of observations. TC-1 was decommissioned and re-entered the atmosphere on October 14, 2007. Contact was lost with TC-2 in early August 2007. The drum-shaped satellites are spin-stabilized, rotating at 15 rpm. They are coated with solar cells, producing 280 W of power. The spacecraft carry two 2.5 m (8.2 ft) experimental rigid booms and two axial telecommunication antenna booms. ESA contributed eight scientific instruments. These are the first-ever European experiments to fly on a Chinese satellite.

SPECIFICATION:

MANUFACTURER: Chinese National Space Administration (CNSA)
LAUNCH DATES: December 29, 2003 (DS-1); July 25, 2004 (DS-2)
ORBITS: DS-1, 570 x 78,970 km (354 x 49,071 mi), 28.5° inclination; DS-2, 690 x 38,230 km (429 x 23,756 mi), 90.1° inclination
LAUNCH MASS: 330 kg (726 lb)
LAUNCH SITES: DS-1, Xichang, China; DS-2, Taiyuan, China
LAUNCHER: Long March 2C
BODY DIMENSIONS: 2.1 x 1.4 m (6.9 x 4.6 ft)
PAYLOAD: Active Spacecraft Potential Control (ASPOC), DS-1 only; Fluxgate Magnetometer (FGM), DS-1 and DS-2; Plasma Electron and Current Experiment (PEACE), DS-1 and DS-2; Neutral Atom Imager (NUADU), DS-2 only; Hot Ion Analyzer (HIA), DS-1 only; Low-Energy Ion Detector (LEID), DS-2 only; Spatio-Temporal Analysis of Field (STAFF)/Digital Wave Processor (DWP), DS-2 only; Low-Frequency Electromagnetic Wave (LFEW), DS-2 only; High-Energy Electron Detector (HEED), DS-1 and DS-2; High-Energy Proton Detector (HEPD), DS-1 and DS-2; Heavy Ion Detector (HID), DS-1 and DS-2

Galileo Orbiter USA

First Jupiter orbiter, first asteroid flybys

Galileo was carried into Earth orbit by *Atlantis* on mission STS-34. It flew past Venus at a range of 16,000 km (9942 mi) on February 10, 1990. It also performed two flybys of Earth.

The first spacecraft flyby of an asteroid took place on October 29, 1991, when Galileo passed 1601 km (995 mi) from Gaspra. On August 28, 1993, it flew past Ida and discovered a small satellite, Dactyl. It also observed the impacts on Jupiter of fragments of Comet Shoemaker-Levy 9. On July 13, 1995, it released the Jupiter descent probe. After adjusting its course, Galileo entered orbit around Jupiter on December 7, 1995.

During its primary mission, Galileo made four flybys of Ganymede, three of Callisto, and three of Europa. A two-year extension included eight close encounters with Europa. The Galileo Millennium Mission included joint studies of Jupiter with Cassini in December 2000. The mission ended on September 21, 2003 when the orbiter was deliberately sent into Jupiter's atmosphere.

Galileo was the first dual-spin planetary spacecraft. A spinning section rotated at about 3 rpm, and a despun section counterrotated to provide a fixed orientation for cameras and other sensors.

SPECIFICATION:

MANUFACTURER: NASA-JPL
LAUNCH DATE: October 18, 1989
JUPITER ORBIT (INITIAL): 286,160 x 17,850,000 km (177,820 x 11,091,780 mi), 1.5° inclination
LAUNCH MASS: 2,223 kg (4,902 lb) excluding Probe
LAUNCH SITE: Cape Canaveral, Florida
LAUNCHER: Shuttle *Atlantis* (STS-30)
BODY DIMENSIONS: 5.3 m (17 ft) high
PAYLOAD: Solid State Imaging (SSI); Near-Infrared Mapping Spectrometer (NIMS); Ultraviolet Spectrometer and Extreme Ultraviolet Spectrometer (UVS/EUVS); Photopolarimeter-Radiometer (PPR); Magnetometer (MAG); Energetic Particles Detector (EPD); Plasma Detector (PLS); Plasma Wave Spectrometer (PWS); Heavy Ion Counter (HIC); Dust Detection System (DDS)

Galileo Probe USA

Jupiter atmospheric probe

SPECIFICATION:

MANUFACTURER: Hughes Aircraft Company
LAUNCH DATE: October 18, 1989
ORBIT: Heliocentric
LAUNCH MASS: 337 kg (741 lb)
LAUNCH SITE: Cape Canaveral, Florida
LAUNCHER: Shuttle *Atlantis* (STS-30)
BODY DIMENSIONS: 1.3 x 0.9 m (4.2 x 3 ft)
PAYLOAD: Atmospheric Structure Instrument (ASI); Neutral Mass Spectrometer (NMS); Helium Abundance Detector (HAD); Net Flux Radiometer (NFR); Nephelometer (NEP); Lightning and Radio Emission Detector (LRD); Energetic particle Investigation (EPI)

Galileo's descent probe, which had been carried from Earth aboard the parent spacecraft, was released on July 13, 1995, and began a five-month free fall toward Jupiter.

The probe included a cone-shaped deceleration module that consisted of an aeroshell and an aft cover; this would slow the descent and protect the probe from frictional heating during atmospheric entry. The descent module carried a radio transmitter and seven scientific instruments to send back data on sunlight and heat flux, pressure, temperature, winds, lightning, and atmospheric composition during its passage into Jupiter's atmosphere.

Atmospheric entry took place on December 7, 1995. The probe transmitted for 57 min, floating down to a depth of about 200 km (125 mi). It survived for 58 min until high temperatures silenced its transmitters. The last data were sent from atmospheric pressure 23 times that of the average air pressure at sea level on Earth. Jupiter's atmosphere was found to be surprisingly dry; this was later explained by the probe having entered a rare "hot spot" in the clouds.

Genesis USA

Solar wind sample return

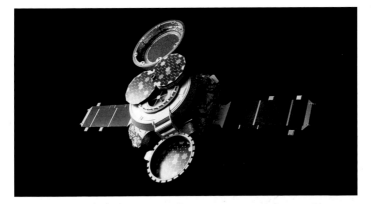

Genesis was the fifth mission in NASA's Discovery program. Its purpose was to trap particles of the solar wind and return them to Earth.

After launch, Genesis traveled to the L1 Lagrange point about 1.5 million km (930,000 mi) from Earth, where it was able to collect 10–20 µg of pristine particles on ultra-pure wafers of gold, sapphire, silicon, and diamond. The collection ended on April 1, 2004. After a complex series of maneuvers and an Earth flyby on May 2, Genesis returned the capsule to Earth on September 8, 2004. The main bus continued into a heliocentric orbit.

The 152 cm (5 ft) diameter return capsule was to be captured in mid-air by a helicopter, but the parachute system failed to deploy. The capsule hit the ground at the Utah landing site at nearly 320 km/h (200 mph). Scientists are now trying to retrieve as many samples as possible.

With its solar arrays extended, the spacecraft resembled an unbuckled wristwatch. A hinged clamshell mechanism opened and closed the sample return capsule in the center. The medium-gain antenna was on the underside, and the low-gain antennas were on the solar panels.

SPECIFICATION:

MANUFACTURER: Lockheed Martin Space Systems
LAUNCH DATE: August 8, 2001
ORBIT: halo at L1 Lagrange point
LAUNCH MASS: 636 kg (1402 lb)
LAUNCH SITE: Cape Canaveral, Florida
LAUNCHER: Delta II 7326
BODY DIMENSIONS: 2.3 x 2 m (7.5 x 6.6 ft)
PAYLOAD: 3 solar wind collector arrays; 2 bulk collector arrays; ion and electron monitors; 2 solar wind concentrators

Geotail JAPAN/USA

Magnetosphere studies

SPECIFICATION:

MANUFACTURER: ISAS
LAUNCH DATE: July 24, 1992
ORBIT: 57,400 x 191,340 km
(35,668 x 118,896 mi), 7.5°
inclination
LAUNCH MASS: 970 kg (2134 lb)
LAUNCH SITE: Cape Canaveral,
Florida
LAUNCHER: Delta II 6925
BODY DIMENSIONS: 2.2 x 1.6 m
(7.2 x 5.2 ft)
PAYLOAD: Electric Field Detector
(EFD); Fluxgate Magnetometer
(MGF); Low-Energy Particles
experiment (LEP);
Comprehensive Plasma
Investigation (CPI); High-Energy
Particles experiment (HEP);
Energetic Particle and Ion
Composition experiment (EPIC);
Plasma Waves Investigation
(PWI)

Geotail was the first satellite launched under the International Solar Terrestrial Physics program. It studies the structure and dynamics of the magnetotail from a highly elliptical orbit. In the first two years, double lunar swingby maneuvers kept apogees on the night side at distances of 80–220 Earth radii (510,240–1,403,160 km or 317,048–871,882 mi). The present orbit is 9 x 30 Earth radii (57,402–191,340 km or 35,668–118,893 mi).

Geotail is cylindrical with body-mounted solar cells. The spin-stabilized cylindrical spacecraft rotates at a rate of 20 rpm with a spin axis nearly perpendicular to the ecliptic and mechanically despun antennas. Geotail has seven instruments—two from the US and five from Japan. Real-time X-band telemetry is received at the Usuda Deep Space Center. The satellite carries two tape recorders, each with a capacity of 450 Mb, which allow 24-hour data coverage.

Geotail is a joint program of the Japanese Institute of Space and Astronautical Science (ISAS, now part of JAXA) and NASA. ISAS developed the spacecraft and provided four of the science instruments. NASA provided the launch and the other instruments. The spacecraft is operated by ISAS but telemetry is received by both agencies.

Hayabusa (Muses-C) JAPAN

Asteroid orbiter/lander/sample return

Hayabusa (Falcon) is a Japanese mission to observe near-Earth asteroid 25143 Itokawa from close range, collect and return a surface sample, and demonstrate various new technologies.

After launch, the ion engines, which use xenon gas, were used to reach Itokawa. Rendez-vous finally occurred in September 2005, with the spacecraft remaining in a station-keeping position 20 km (12.4 mi) from the asteroid. On October 3, 2005, Hayabusa lost its Y-axis reaction wheel, leaving one reaction wheel and two thrusters to maintain attitude control.

After two aborted attempts, Hayabusa briefly landed twice on Itokawa. On November 25, 2005, it may have fired two bullets that enabled fragments to be collected for analysis on Earth. On December 9, contact was lost after a major thruster propellant leak. Contact was restored in March 2006, and the spacecraft set off for Earth on April 25, 2007, with only one of four ion engines working properly. Arrival at Earth is expected in 2010, after two orbits of the Sun, with the sample capsule to be retrieved near Woomera, Australia.

SPECIFICATION:

MANUFACTURER: ISAS
LAUNCH DATE: May 9, 2003
ORBIT: Heliocentric, asteroid orbits (variable)
LAUNCH MASS: 530 kg (1168 lb)
LAUNCH SITE: Kagoshima, Japan
LAUNCHER: M-V-5
BODY DIMENSIONS: 1.5 x 1.5 x 1.1 m (4.9 x 4.9 x 3.4 ft)
PAYLOAD: Asteroid Multi-Band Imaging Camera (AMICA); Light Detection and Ranging Instrument (LIDAR); Near-Infrared Spectrometer (NIRS); X-Ray Fluorescence Spectrometer (XRS)

Hinode (Solar-B) JAPAN/USA/UK

Solar observatory

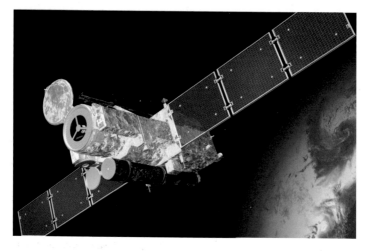

Hinode (Sunrise) is a collaborative mission led by the JAXA with participation by ESA, the UK, and the USA. Its mission is to study the Sun's magnetic field and explosive energy.

The spacecraft is built around the optical telescope assembly. Hinode is three-axis stabilized and points continuously toward the Sun. Attitude is sensed by an inertial reference unit and by gyros with a combination of two Sun sensors and a star tracker, providing absolute calibration.

The National Astronomical Observatory of Japan (NAOJ) developed the SOT, which provides detailed views of the Sun's lower atmosphere and is the first space telescope to measure the Sun's three-dimensional magnetic field vector. NAOJ also developed the SXT—the highest-resolution X-ray telescope ever flown in space—in collaboration with the US Smithsonian Astrophysical Observatory. The SXT is mounted on the opposite side of the SOT from the EIS, which was built by a UK-led consortium.

SPECIFICATION:

MANUFACTURER: Mitsubishi Electric Corporation (MELCO)
LAUNCH DATE: September 23, 2006
ORBIT: 674 x 695 km (420 x 435 mi), 98.1° inclination (Sun-synchronous)
LAUNCH MASS: 870 kg (1910 lb)
LAUNCH SITE: Uchinoura, Kagoshima, Japan
LAUNCHER: M-V-7
BODY DIMENSIONS: 1.6 x 1.6 x 1.4 m (5.2 x 5.2 x 4.6 ft)
PAYLOAD: Solar Optical Telescope (SOT); Extreme-Ultraviolet Imaging Spectrometer (EIS); Solar X-Ray Telescope (SXT)

Huygens EUROPE

Titan lander

Huygens was an entry probe designed to descend through the thick, nitrogen-rich atmosphere of Saturn's moon Titan. The ESA spacecraft, named after Dutch astronomer Christiaan Huygens, was carried to Saturn on board NASA's Cassini orbiter.

A design error in the radio receiver, discovered en route, meant that the arrival date at Titan had to be delayed by seven weeks in order to reduce the Doppler shift in the Huygens signal during descent. Huygens was released towards Titan on December 25, 2004, and coasted for 20 days before entering Titan's atmosphere on January 14, 2005. As the probe descended by parachute to the surface, its instruments relayed data to the Cassini orbiter as it passed 60,000 km (37,282 mi) overhead.

After separation from Cassini, power was provided by five lithium batteries. Touchdown speed on Titan was about 5 m/s (16.4 ft/s). The descent phase lasted 2 h 27 min, with a further 1 h 10 min on the surface. Some of the descent data were lost because one of the two transmission channels was never turned on due to a configuration error.

SPECIFICATION:

MANUFACTURER: Aerospatiale
LAUNCH DATE: October 15, 1997
ORBIT: Heliocentric
LAUNCH MASS: 348.3 kg (766 lb)
LAUNCH SITE: Cape Canaveral, Florida
LAUNCHER: Titan IVB-Centaur
BODY DIMENSIONS: 2.8 m (9 ft) diameter
PAYLOAD: Aerosol Collector and Pyrolyzer (ACP); Descent Imager/Spectral Radiometer (DISR); Doppler Wind Experiment (DWE); Gas Chromatograph and Mass Spectrometer (GCMS); Huygens Atmosphere Structure Instrument (HASI); Surface Science Package (SSP)

Lunar Reconnaissance Orbiter/Lunar Crater Observation and Sensing Satellite (LRO/LCROSS) USA

Lunar orbiter and impactor

SPECIFICATION:

MANUFACTURER: NASA-Goddard (LRO); Northrop Grumman Space Technologies (LCROSS)
LAUNCH DATE: October 31, 2008?
ORBIT: Lunar, 50 km (31 mi) circular, polar
LAUNCH MASS: LRO, 1706 kg (3753 lb); LCROSS Shepherd, 534 kg (1177 lb)
LAUNCH SITE: Cape Canaveral, Florida
LAUNCHER: Atlas V 401
BODY DIMENSIONS: 2 x 2 x 3 m (6.6 x 6.6 x 9.8 ft)
PAYLOAD:
LRO—Cosmic Ray Telescope for the Effects of Radiation (CRaTER); Diviner Lunar Radiometer Experiment (DLRE); Lunar Exploration Neutron Detector (LEND); Lunar Orbiter Laser Altimeter (LOLA); Lyman-Alpha Mapping Project (LAMP); Lunar Reconnaissance Orbiter Camera (LROC)
LCROSS—Visible context cameras; visible total luminance diode (photometer); 2 mid-infrared cameras; 2 near-Infrared cameras; 1 visible spectrometer; 2 near-infrared spectrometers

NASA's LRO/LCROSS will map the lunar surface to help pave the way for extended human surface exploration. It will take approximately four days to reach the Moon and it will then maneuver into a low operational orbit and spend a year mapping the surface and searching for polar ice pockets in shaded craters, which may be of use to future human explorers.

The LCROSS Shepherd spacecraft performs a final targeting maneuver 86 days after launch to line up the LRO Centaur upper stage for impact. The Shepherd spacecraft separates from Centaur and performs a delay burn that leaves it 10 min behind the Centaur. Shepherd then observes the plume when the Centaur impacts. Fifteen minutes later, Shepherd will also impact, creating a smaller, Earth-observable plume. The LCROSS spacecraft structure is based on the EELV Secondary Payload Adaptor ring that contains avionics, a small propulsion system, a single-panel body-mounted solar array, two S-Band omni antennas, and two medium-gain horns. Above the Adaptor ring is the LRO, a three-axis stabilized, nadir-pointed spacecraft.

Magellan USA

Venus orbiter

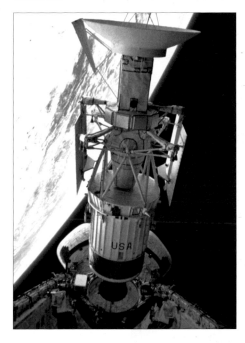

SPECIFICATION:

MANUFACTURER: Martin Marietta
Astronautics
LAUNCH DATE: May 4, 1989
INITIAL VENUS ORBIT: 297 x
8463 km (185 x 5259 mi),
85.5° inclination
LAUNCH MASS: 3445 kg
LAUNCH SITE: Kennedy Space
Center, Florida
LAUNCHER: Shuttle *Atlantis*
(STS-30)
DIMENSIONS: 6.4 x 9.2 m
(21 x 30.2 ft) with solar arrays
deployed
PAYLOAD: Synthetic aperture
radar

Magellan was designed to conduct the most comprehensive observation of the surface and gravitational features of Venus ever undertaken. The spacecraft was topped by a 3.7 m (12 ft) diameter dish antenna (from *Voyager*), which acquired imaging and radiometry data and communicated with Earth.

For 37 min of each orbit, the radar imaged a 24 km (15 mi) wide swathe of the surface while acquiring altimetry and radiometry data.

Lifted into Earth orbit by the Shuttle, Magellan was launched on its 15-month journey to Venus by an Inertial Upper Stage (IUS) booster. It arrived in Venus's orbit on August 10, 1990. During its 243-day (one Venus rotation) primary mission, it mapped 83.7% of the surface at resolutions of between 120 and 360 m (394 and 1180 ft).

After five complete orbital cycles, Magellan eventually mapped 98% of Venus and obtained repeat coverage of many areas. After several adjustments to lower orbits, including a "windmill experiment" in Venus's outer atmosphere, it was deliberately deorbited on October 13, 1994. By then it had returned more data than all previous NASA planetary missions.

Mars Exploration Rovers (Spirit and Opportunity/ MER-A and MER-B) USA

Two Mars landers

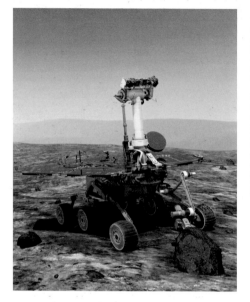

SPECIFICATION:

MANUFACTURER: NASA-JPL
LAUNCH DATES: Spirit, June 10,
2003; Opportunity, July 7, 2003
ORBIT: Heliocentric, landing
LAUNCH MASS: 1062 kg (2341 lb)
LAUNCH SITE: Cape Canaveral,
Florida
LAUNCHERS: Spirit, Delta II 7925;
Opportunity, Delta II 7925H
ROVER DIMENSIONS: 1.5 x 2.3 x
1.6 m (4.9 x 7.5 x 5.2 ft)
PAYLOAD: 2 Panoramic cameras;
miniature thermal emission
spectrometer; Mössbauer
spectrometer; alpha particle
X-ray spectrometer; microscopic
imager; rock abrasion tool;
3 magnet arrays

The Exploration Rovers were designed to spend at least three months searching for evidence of water on Mars. The lander was equipped with a downward-looking camera and a radar.

Each six-wheeled rover has a deck of solar panels, three antennas, an elevated panoramic camera, and an arm carrying most of the science instruments. The electronics box is warmed by electrical heaters and eight radioisotope heaters. The rocker-bogie suspension system bends at its joints. Navigation software and hazard-avoiding capabilities enable the rover to make its own way toward a destination uploaded to the computer.

Spirit landed on the plains of Gusev crater (January 4, 2004). Opportunity landed inside a small impact crater on Meridiani Planum (January 25). As of June 2007, both rovers were still operational. Spirit climbed to the top of the Columbia Hills, 106 m (348 ft) above the plains. One of its wheels was no longer operating. As of sol 1225 (June 14, 2007), Spirit had traveled 7.1 km (4.4 mi). Opportunity had explored several craters; its total distance as of sol 1204 (June 14, 2007) was 11.4 km (7.1 mi).

Mars Express

Mars orbiter

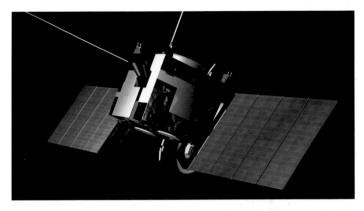

Mars Express comprised the Mars Express Orbiter and the Beagle 2 lander (mainly funded by the UK). The 68 kg (150 lb) Beagle 2 was shaped like a closed clam and would open out to expose its solar panels after landing. Its main purpose was to search for evidence of life—past or present—on Mars. It was released from Mars Express on December 19, 2003, and made a ballistic entry into the Martian atmosphere on December 25. Beagle 2 was expected to land on Isidis Planitia after deploying a parachute and gas bags. No signals were received, however, and the lander was declared lost.

The orbiter went into a 250 x 150,000 km (155 x 93,208 mi) orbit on December 25. Four engine firings changed this to the near-polar operational orbit. Concerns about deployment of the radar experiment's long, whip-like antennas delayed the start of radar sounding for nearly two years. The radar started its science operations on July 4, 2005, and sent back the first subsurface soundings of Mars. The mission has been extended until May 2009.

SPECIFICATION:

MANUFACTURER: EADS Astrium
LAUNCH DATE: June 2, 2003
MARS ORBIT: 258 x 11,560 km (160 x 7183 mi), 86.3° inclination
LAUNCH MASS: 1123 kg (2470 lb)
LAUNCH SITE: Baikonur, Kazakhstan
LAUNCHER: Soyuz-Fregat
BODY DIMENSIONS: 1.5 x 1.8 x 1.4 m (4.9 x 5.9 x 4.6 ft)
PAYLOADS: Mars Express—High-Resolution Stereoscopic Camera (HRSC); Infrared Mineralogical Mapping Spectrometer (OMEGA); Planetary Fourier (infrared) Spectrometer (PFS); Spectroscopic Investigation of the Characteristics of the Atmosphere of Mars (SPICAM); Analyzer of Space Plasma and Energetic Atoms (ASPERA); Mars Advanced Radar for Subsurface and Ionospheric Sounding (MARSIS)
Beagle 2—Gas Analysis Package; environmental sensors; 2 stereo cameras; Mössbauer spectrometer; X-ray spectrometer; rock corer/grinder; microscope; mole

Mars Global Surveyor (MGS) USA

Mars orbiter

MGS was the first in a new series of NASA Mars spacecraft. Its task was to study the entire Martian surface, atmosphere, and interior over one Martian year—nearly two Earth years.

MGS was boosted from Earth parking orbit by the third-stage Star 48B solid rocket. Its main engine fired for Mars orbit insertion on September 12, 1997. One of the solar panels failed to latch properly when it was deployed. A new schedule involved slower aerobraking, which finally ended in February 1999. MGS was then in a circular polar science mapping orbit. The side of the bus with the remote sensing instruments always faced the Martian surface. From launch through aerobraking, the high-gain antenna remained fixed to one side of the spacecraft, and the spacecraft had to be slewed for communications with Earth. The "official" mapping mission began on March 9, 1999, and continued until November 2, 2006, when communications were lost. MGS operated longer at Mars than any other spacecraft in history. Its camera returned more than 240,000 images.

SPECIFICATION:

MANUFACTURER: Lockheed Martin Astronautics
LAUNCH DATE: November 7, 1996
INITIAL MARS ORBIT: 258 x 54,021 km (160 x 33,563 mi)
LAUNCH MASS: 1062 kg (2336 lb)
LAUNCH SITE: Cape Canaveral, Florida
LAUNCHER: Delta II 7925
BODY DIMENSIONS: 1.2 x 1.2 x 1.8 m (3.9 x 3.9 x 5.9 ft)
PAYLOAD: Mars Orbiter Camera (MOC); Mars Orbiter Laser Altimeter (MOLA); Thermal Emission Spectrometer (TES); Magnetometer/Electron Reflectometer (MAG/ER); Mars Relay Antenna (MR)

Mars Odyssey <small>USA</small>

Mars orbiter

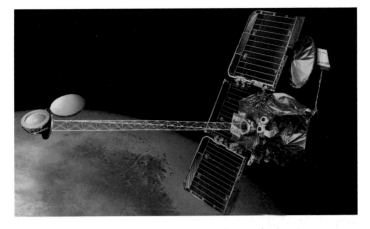

Mars Odyssey is the remaining part of the Mars Surveyor 2001 Project, which originally consisted of the Mars Surveyor 2001 Orbiter and the Mars Surveyor 2001 Lander. Odyssey is making global observations of Mars to improve our understanding of the planet's climate and geologic history.

The propulsion module contains the main engine, fuel tanks, and thrusters. The equipment module supports engineering components, the radiation experiment, and a science deck.

Odyssey entered orbit around Mars on October 24, 2001, and reached its operational mapping orbit on January 30. The science mapping mission began on February 19, 2002, in a 400 km (249 mi) circular, polar, nearly Sun-synchronous orbit. After only 10 days, the gamma-ray instruments recorded significant quantities of hydrogen close to the surface of Mars—evidence of underground water ice.

The primary science mission continued through August 2004. The second extended mission is due to end in October 2008. Odyssey also serves as the primary communications relay for NASA's Mars Exploration Rovers.

SPECIFICATION:

MANUFACTURER: Lockheed Martin Astronautics
LAUNCH DATE: April 7, 2001
OPERATIONAL MARS ORBIT: 400 km (249 mi) circular, 93.2° inclination
LAUNCH MASS: 729.7 kg (1608 lb)
LAUNCH SITE: Cape Canaveral, Florida
LAUNCHER: Delta II 7925
BODY DIMENSIONS: 2.2 x 1.7 x 2.6 m (7.2 x 5.6 x 8.5 ft)
PAYLOAD: Thermal Emission Imaging System (THEMIS); Gamma-Ray Spectrometer/neutron spectrometer/high-energy neutron detector (GRS); Martian Radiation Environment experiment (MARIE)

Mars Pathfinder USA

Mars lander/rover

Mars Pathfinder was the second mission in NASA's Discovery program of "faster, better, cheaper" missions and the first to place a NASA automated rover on another world. Its main objective was to demonstrate a new method of landing on Mars using a parachute and airbags.

The spacecraft made a ballistic entry into the Martian atmosphere on July 4, 1997, and landed with the aid of parachutes, rockets, and airbags, taking atmospheric measurements on the way.

The lander carried a camera on a mast to survey its surroundings and atmospheric sensors on one of the solar panels. Sojourner was 0.28 m high, 0.6 m long, and 0.5 m wide (0.9 x 2.1 x 1.6 ft), with a ground clearance of 0.1 m (0.4 ft).

On arrival, Pathfinder bounced about 15 m (50 ft) into the air, bouncing another 15 times and rolling before coming to rest about 1 km (0.6 mi) from the initial impact site. Sojourner drove for 83 Martian days, until a flat battery on the lander halted communications on September 27. The rover cameras took 550 images, and 16 chemical analyses of rocks and soil were made by the X-ray spectrometer.

SPECIFICATION:

MANUFACTURER: Jet Propulsion Laboratory
LAUNCH DATE: December 4, 1996
ORBIT: Heliocentric, landing
LAUNCH MASS (LANDER): 895 kg (1973 lb)
LAUNCH SITE: Cape Canaveral, Florida
LAUNCHER: Delta II 7925
BODY DIMENSIONS (LANDER): 0.9 m (3 ft) tall (hexagonal)
PAYLOAD: Lander—Imaging system (IMP), included magnets and windsocks; Atmospheric Structure/Meteorology package (ASI/MET)
Sojourner rover—Imaging system (two forward black-and-white cameras, one rear color camera); Alpha/Proton/X-ray Spectrometer (APXS); Materials Adherence Experiment (MAE)

Mars Reconnaissance Orbiter (MRO) USA

Mars orbiter

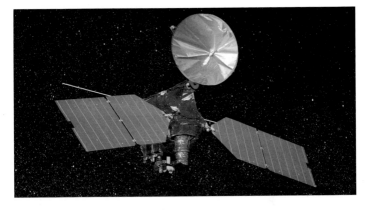

MRO is the largest US spacecraft ever sent to Mars and the first launch of an interplanetary mission on an Atlas V. The spacecraft is designed to characterize the surface, subsurface, and atmosphere of Mars, particularly the history of water, past and present.

The box-shaped orbiter carries its science instruments on one side. Each of the two solar arrays is 5 m (17.6 ft) long by 2.5 m (8.3 ft) wide. Output at Mars is 2000 W. The solid-state recorder provides storage space for up to 160 Gb of data.

The Hi-RISE camera obtains high-resolution images with stereo coverage. MRO reached Mars on March 10, 2006. During six months of aerobraking it dipped into the upper atmosphere 426 times, gradually decreasing the highest point of its elliptical orbit from 45,000 km (28,000 mi) to 486 km (302 mi). The main science mission began in November 2006 and will last for one Mars year (687 Earth days). MRO will also act as a relay for NASA landers/rovers. The prime mission is scheduled to end on December 31, 2010. MRO will provide more science data than all previous Mars missions combined.

SPECIFICATION:

MANUFACTURER: Lockheed Martin Space Systems
LAUNCH DATE: August 12, 2005
OPERATIONAL MARS ORBIT: 250 km x 316 km (155 x 196 mi), 92.65° inclination, sun-synchronous
LAUNCH MASS: 2180 kg (4796 lb)
LAUNCH SITE: Cape Canaveral, Florida
LAUNCHER: Atlas V-401 (with Centaur upper stage)
BODY DIMENSIONS: 6.5 x 13.6 m (21 x 45 ft)
PAYLOAD: High-Resolution Imaging Science Experiment (Hi-RISE); Mars Color Imager (MARCI); Compact Reconnaissance Imaging Spectrometer for Mars (CRISM); context camera; Shallow Radar (SHARAD); Mars Climate Sounder (MCS); optical navigation camera; Electra UHF radio relay; Ka-band telecommunications demonstration

MESSENGER (MErcury Surface, Space ENvironment, GEochemistry, and Ranging) USA

Mercury orbiter

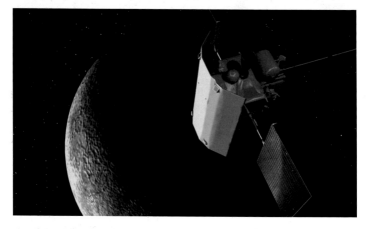

MESSENGER, the seventh mission in NASA's Discovery program, will be the first spacecraft to orbit Mercury and image its entire surface. Its 7.9 billion km (4.9 billion mi) journey will include one Earth flyby (August 2, 2005), two Venus flybys (October 24, 2006, and June 6, 2007), and three Mercury flybys (January 14 and October 6, 2008, and September 30, 2009), with orbit insertion around Mercury on March 18, 2011.

The spacecraft has a graphite composite structure—strong, lightweight and heat tolerant—and carries a front-mounted, ceramic-fabric sunshade. It has one bipropellant thruster for large maneuvers and 16 thrusters for attitude control.

MESSENGER will follow an elliptical orbit around Mercury, with a low point at 60° north. Its prime mission will last for one Earth year (four Mercury years). It will search for water ice in polar craters and collect data on the composition and structure of Mercury's crust and iron core, its topography and geologic history, and the nature of its thin atmosphere and magnetosphere.

SPECIFICATION:

MANUFACTURER: Johns Hopkins University Applied Physics Laboratory
LAUNCH DATE: August 3, 2004
MERCURY ORBIT: 200 x 15,193 km (124 x 9439 mi), 80° inclination
LAUNCH MASS: 1100 kg (2424 lb)
LAUNCH SITE: Cape Canaveral, Florida
LAUNCHER: Delta II 7925H
BODY DIMENSIONS: 1.4 x 1.9 x 1.3 m (4.7 x 6.1 x 4.2 ft)
PAYLOAD: Mercury Dual Imaging System (MDIS); Gamma-Ray and Neutron Spectrometer (GRNS); X-Ray Spectrometer (XRS); Energetic Particle and Plasma Spectrometer (EPPS); Mercury Atmospheric/Surface Composition Spectrometer (MASCS); Mercury Laser Altimeter (MLA); Magnetometer (MAG)

NEAR Shoemaker (Near Earth Asteroid Rendezvous) USA

Asteroid orbiter/lander

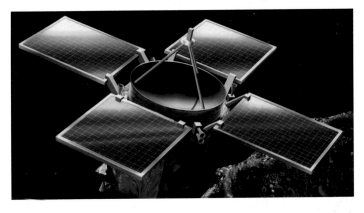

NEAR Shoemaker was the first spacecraft to be launched in NASA's Discovery program and the first to rely on solar cells for power beyond Mars orbit. After launch, the spacecraft was renamed in honor of planetary geologist Gene Shoemaker.

En route to Eros, it passed within 1200 km (745 mi) of Mathilde on June 27, 1997, and discovered that the heavily cratered main belt asteroid is a low-density rubble pile. Following an Earth gravity assist on January 23, 1998, it headed towards Eros, but entry into orbit was postponed after an aborted engine burn on December 20. Observations of Eros were made during a flyby at a range of 3827 km (2378 mi) on December 23.

The second attempt to enter orbit was successful on February 14, 2000. The spacecraft circled the asteroid for a year, dipping to within 2.7 km (1.7 mi) of the surface. After collecting 10 times the data initially expected, it made the first landing on an asteroid. The touchdown took place on February 12, 2001, at a speed of 1.6 m/s (5.2 ft/s). Its gamma-ray spectrometer continued to return information for two weeks, until contact was lost on March 1, 2001.

SPECIFICATION:

MANUFACTURER: Johns Hopkins University Applied Physics Laboratory
LAUNCH DATE: February 17, 1996
INITIAL ASTEROID ORBIT: 321 x 366 km (200 x 227 mi)
LAUNCH MASS: 805 kg (1771 lb)
LAUNCH SITE: Cape Canaveral, Florida
LAUNCHER: Delta II 7925
BODY DIMENSIONS: 1.7 x 1.7 x 1.7 m (5.6 x 5.6 x 5.6 ft)
PAYLOAD: Multispectral imager; Magnetometer (MAG); Near-infrared spectrometer; X-ray/gamma ray spectrometer; laser rangefinder

New Horizons   USA

Jupiter/Pluto/Kuiper Belt flybys

New Horizons, the first spacecraft to be developed under NASA's New Frontiers program, will make the first flybys of Pluto and Charon and at least one Kuiper belt object. It is the fastest spacecraft ever launched, crossing lunar orbit in 9 h and passing Jupiter 13 months later.

Powered by a single radioisotope thermo-electric generator, each of the onboard science instruments uses between 2 and 10 W. Data are recorded on one of two onboard solid-state memory banks before playback to Earth. The spacecraft communicates with Earth through a 2.1 m (6.9 ft) high-gain antenna. The Jupiter gravity assist on February 28, 2007, reduced the journey time by five years. The spacecraft will make a flyby of Pluto and Charon in July 2015 at a speed of 69,200 km/h (43,000 mph), characterizing the global geology of Pluto and Charon, mapping surface compositions and temperatures, and studying Pluto's sparse atmosphere. It will then visit one or more of the icy, primordial bodies in the Kuiper belt.

SPECIFICATION:

MANUFACTURER: Johns Hopkins University Applied Physics Laboratory
LAUNCH DATE: January 19, 2006
ORBIT: Escape
LAUNCH MASS: 465 kg (1025 lb)
LAUNCH SITE: Cape Canaveral, Florida
LAUNCHER: Atlas V-551 (Centaur upper stage and STAR-48B third stage)
BODY DIMENSIONS: 0.7 x 2.1 x 2.7 m (2.3 x 6.9 x 9.0 ft)
PAYLOAD: ALICE ultraviolet imaging spectrometer; Long Range Reconnaissance Imager (LORRI); RALPH imaging system; Solar Wind at Pluto (SWAP); Pluto Energetic Particle Spectro-meter Science Investigation (PEPSSI); Student Dust Counter (SDC)

Phoenix USA

Mars lander

Phoenix is the first of NASA's low-cost Mars Scout missions. It uses many components of two unsuccessful Mars missions. Phoenix is designed to study the history of water and search for complex organic molecules in the ice-rich soil of the Martian Arctic.

During a ballistic atmospheric entry, the spacecraft will be protected by an aeroshell. As its speed drops, the aeroshell is jettisoned and a parachute is deployed. Descent engines fire as the lander is released from its parachute and backshell, about 570 m (1900 ft) above the surface. A soft touchdown is made using controlled bursts from 12 hydrazine thrusters. The scheduled arrival date is May 25, 2008, at a site near the Martian Arctic Circle.

During its 90-day primary mission, Phoenix will dig trenches with its robotic arm in the frozen layers of water below the surface. Samples will be analyzed in eight ovens and four cells where water from Earth is mixed with Martian soil. All data will be relayed through the Mars Reconnaissance Orbiter and Mars Odyssey.

SPECIFICATION:

MANUFACTURER: Lockheed Martin Space Systems
LAUNCH DATE: August 4, 2007
ORBIT: Heliocentric, lander
LAUNCH MASS: 670 kg (1477 lb)
LAUNCH SITE: Cape Canaveral, Florida
LAUNCHER: Delta II 7925
LANDER DIMENSIONS: 2.2 x 5.5 m (7.2 x 18.1 ft) with solar arrays and mast deployed
PAYLOAD: Microscopy, Electrochemistry, and Conductivity Analyzer (MECA); Robotic Arm Camera (RAC); Surface Stereo Imager (SSI); Thermal and Evolved Gas Analyzer (TEGA); Mars Descent Imager (MARDI); Meteorological Station (MET)

Polar USA

Magnetosphere/aurora studies

SPECIFICATION:

MANUFACTURER: Martin Marietta (now Lockheed Martin)
LAUNCH DATE: February 24, 1996
ORBIT: 11,500 x 57,000 km (7146 x 35,419 mi), 86° inclination (variable)
LAUNCH MASS: 1250 kg (2750 lb)
LAUNCH SITE: Vandenberg, California
LAUNCHER: Delta II 7925
BODY DIMENSIONS: 2.4 x 2.1 m (7.9 x 6.9 ft)
PAYLOAD: Plasma Waves Investigation (PWI); Magnetic Fields Experiment (MFE); Toroidal Imaging Mass-Angle Spectrograph (TIMAS); Electric Fields Investigation (EFI); Thermal Ion Dynamics Experiment (TIDE); Ultraviolet Imager (UVI); Visible Imaging System (VIS); Polar Ionospheric X-Ray Imaging Experiment (PIXIE); Charge and Mass Magnetospheric Ion Composition Experiment (CAMMICE); Comprehensive Energetic-Particle Pitch-Angle Distribution, Source/Loss Cone Energetic-Particle Spectrometer (CEPPAD/SEPS)

Polar is one of four spacecraft in the Global Geospace Science program, part of the International Solar Terrestrial Physics (ISTP) program. Three of Polar's 12 scientific instruments provide multi-wavelength imaging of the aurora; the others measure the fluxes of charged particles, electrons, and protons in near-Earth space.

Polar is cylindrical, with a despun instrument platform for the imagers and body-mounted solar cells. There are two antennae top and bottom for the EFI experiment, four wire antennae for electric-field and plasma-wave investigations, and two 6 m (19.7 ft) booms for magnetic-field instruments.

Polar was launched to observe the polar magnetosphere and, as its orbit has precessed with time, it has observed the equatorial inner magnetosphere and is now carrying out an extended period of southern-hemisphere coverage. It provided the first-ever footage of auroras simultaneously visible around both polar regions. Polar was designed for a lifetime of three to five years, but it is still operational.

RHESSI (Ramaty High-Energy Solar Spectroscopic Imager) USA

RHESSI is NASA's sixth Small Explorer mission. It was renamed after launch in honor of Reuven Ramaty, a Romanian-born NASA scientist.

RHESSI's primary mission is to explore the basic physics of particle acceleration and explosive energy release in solar flares. RHESSI watches the Sun in X-rays and gamma rays and is the first spacecraft to make high-resolution images of flares using their high-energy radiation. The mission was expected to last two to three years but has been extended until at least April 2008.

RHESSI is spin-stabilized and orbits Earth about 15 times a day. Four 5.7 m (18.7 ft) long solar arrays provide 440 W of power. The imaging spectrometer has the best angular and spectral resolution of any hard X-ray or gamma-ray instrument flown in space. Its imaging capability is achieved with fine tungsten and/or molybdenum grids that modulate the solar X-ray flux as the spacecraft rotates at about 15 rpm. Up to 20 detailed images can be obtained per second.

SPECIFICATION:

MANUFACTURER: Spectrum Astro
LAUNCH DATE: February 5, 2002
ORBIT: 600 km (373 mi), 38° inclination
LAUNCH MASS: 293 kg (645 lb)
LAUNCH SITE: Cape Canaveral, Florida (air launched)
LAUNCHER: Pegasus XL
BODY DIMENSIONS: 2.2 x 5.8 m (7.1 x 18.9 ft)
PAYLOAD: Imaging spectrometer

Rosetta/Philae EUROPE

Comet orbiter/lander

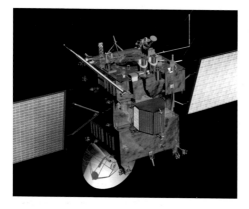

SPECIFICATION:

MANUFACTURER: EADS Astrium
LAUNCH DATE: March 2, 2004
ORBIT: Heliocentric, comet orbits
(variable)
LAUNCH MASS: 3065 kg (6743 lb)
LAUNCH SITE: Kourou, French
Guiana
LAUNCHER: Ariane 5G+
BODY DIMENSIONS: 2.8 x 2.1 x
2.0 m (9.2 x 6.9 x 6.6 ft)
PAYLOAD: Orbiter—Ultraviolet
Imaging Spectrometer (ALICE);
Comet Nucleus Sounding
Experiment by Radio-Wave
Transmission (CONSERT);
Cometary Secondary Ion Mass
Analyzer (COSIMA); Grain Impact
Analyzer and Dust Accumulator
(GIADA); Micro-Imaging Dust
Analysis System (MIDAS);
Microwave Instrument for the
Rosetta Orbiter (MIRO); Optical,
Spectroscopic, and Infrared
Remote Imaging System
(OSIRIS); Rosetta Orbiter
Spectrometer for Ion and Neutral
Analysis (ROSINA); Visible and
Infrared Thermal Imaging
Spectrometer (VIRTIS); Rosetta
Plasma Consortium (RPC), 6
elements
Lander—Modulus PTOLEMY
evolved gas analyzer; Alpha
Particle X-ray Spectrometer
(APXS); Panoramic and
Microscopic Imaging System
(ÇIVA); Cometary Sampling and
Composition experiment
(COSAC); Comet Nucleus
Sounding Experiment by
Microwave Transmission
(CONCERT); Multi-Purpose
Sensors for Surface and
Subsurface Science (MUPUS);
Surface Electrical, Seismic, and
Acoustic Monitoring
Experiments (SESAME); Rosetta
Lander Imaging System (ROLIS);
Rosetta Lander Magnetometer
and Plasma Monitor (ROMAP);
Sample and Distribution Device
(SD2)

Rosetta is shaped like a cube. Its upper part carries 11 science instruments, while the bottom section holds the subsystems. Rosetta was the first mission to use an Ariane 5 for launch into escape orbit. The EPS upper stage boosted Rosetta out of Earth parking orbit.

The 10-year voyage to Comet 67P/Churyumov-Gerasimenko involves three Earth flybys (March 4, 2005; November 13, 2007; and November 13, 2009) and one Mars flyby (February 25, 2007). There will be two flybys of main belt asteroids—2867 Steins (September 5, 2008) and 21 Lutetia (July 10, 2010). Orbit insertion around the comet will take place in August 2014. Rosetta will remain in orbit until after perihelion, observing changes in the comet. The mission will end in December 2015.

In November 2014, the 100 kg (220 lb) Philae will be released at an altitude of about 1 km (0.6 mi) to make the first soft landing on a comet nucleus. Philae has nine experiments and a Sampling Drilling and Distribution device which will drill over 20 cm (8 in) into the surface, collect samples, and deposit them in ovens or deliver them for microscope inspection. Data will be relayed to Earth via the orbiter.

SELENE (Selenological and Engineering Explorer, Kaguya) JAPAN

Lunar orbiter and subsatellites

SPECIFICATION:

MANUFACTURER: JAXA
LAUNCH DATE: September 14, 2007
OPERATIONAL LUNAR ORBIT:
100 km (62 mi) circular, polar
LAUNCH MASS: 2885 kg (6347 lb)
LAUNCH SITE: Tanegashima, Japan
LAUNCHER: H-IIA 2022
BODY DIMENSIONS: 2.1 x 2.1 x
4.8 m (6.9 x 6.9 x 15.7 ft)
ORBITER PAYLOAD: High-
definition television system;
X-ray spectrometer; gamma-ray
spectrometer; multi-band imager;
spectral profiler; terrain camera;
laser altimeter; lunar radar
sounder; charged particle spectrometer; Plasma energy Angle
and Composition Experiment;
Upper-atmosphere and plasma
imager; lunar magnetometer

SELENE (also named Kaguya after a legendary princess) consists of a main orbiter and two 53 kg (117 lb) subsatellites: Rstar (Relay Satellite) and VRAD (Very Long Baseline Interferometry Radio Satellite). The orbiter main bus is box shaped, with an upper module, containing most of the scientific instruments, and a lower propulsion module. A solar array is mounted on one side of the spacecraft. A 1.3 m (4.3 ft) high-gain antenna is mounted 90° from the solar panel. A 12 m (39 ft) magnetometer boom juts out of the top of the spacecraft, and four 15 m (49 ft) radar sounder antennas protrude from the top and bottom corners of the upper module.

The subsatellites are spin-stabilized at 10 rpm and have no propulsion units. Power is provided by a 70 W silicon solar-cell array covering their sides. Both carry one X-band and three S-band radio sources, enabling differential VLBI observations from the ground. The Rstar will relay the Doppler ranging signal between the orbiter and the ground station for the first direct measurement of the gravity field on the lunar far side. The radio signal from the VRAD is also used to detect the lunar ionosphere.

Selene was placed in a highly elliptical parking orbit. After several orbital maneuvers, it was expected to arrive at the Moon in mid-October. The Rstar will be released into a 100 x 2400 km (62 x 1491 mi) polar orbit and the VRAD will be released into a 100 x 800 km (62 x 497 mi) orbit. The main orbiter will operate for one year, using correction burns roughly every two months to maintain the orbit. The orbit may then be lowered to 40–70 km.

SMART-1 (Small Mission for Advanced Research in Technology) EUROPE

Lunar orbiter and technology demonstration

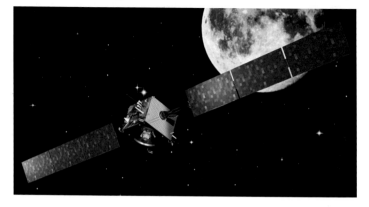

SMART-1 was the first in a proposed series of Small Missions for Advanced Research in Technology. It was primarily a demonstration mission for new instruments and techniques.

The box-shaped orbiter was built around an experimental solar-electric propulsion system that accelerated xenon ions. It also tested new autonomous navigation and space-communication techniques and several experimental science instruments, including use of very short (Ka-band) radio waves and a laser beam for communication with Earth.

Spiralling outward for 14 months, SMART-1 entered the Moon's gravitational field and was captured into a polar orbit on November 15, 2004, after 331 ever-expanding orbits of Earth.

The science mission officially began on February 25, 2005, allowing six months of lunar observations. This was later extended to August 2006. The mission ended on September 3, 2006, with a controlled lunar impact, during SMART-1's 2890th orbit of the Moon. Ground-based telescopes observed a bright flash from the impact.

SPECIFICATION:

MANUFACTURER: Swedish Space Corporation
LAUNCH DATE: September 27, 2003
OPERATIONAL MOON ORBIT: 471 x 2880 km (293 x 1790 mi), 90.06° inclination
LAUNCH MASS: 370 kg (814 lb)
LAUNCH SITE: Kourou, French Guiana
LAUNCHER: Ariane 5
BODY DIMENSIONS: 1.2 x 1.2 x 0.9 m (3.8 x 3.8 x 3.1 ft)
PAYLOAD: Advanced Moon Imaging Experiment (AMIE); Demonstration Compact Imaging X-ray Spectrometer/X-ray Solar Monitor (D-CIXS/XSM); Spacecraft Potential, Electron and Dust Experiment (SPEDE); SMART-1 Infrared Spectrometer (SIR); Ka-band TT&C Experiment (KaTE); Onboard Autonomous Navigation (OBAN); LaserLink; Electric Propulsion Diagnostics Package (EPDP)

SOHO (Solar and Heliospheric Observatory) EUROPE/USA

Solar/solar wind observatory

SOHO moves around the Sun in step with the Earth, slowly orbiting around the L1 Lagrange point, 1.5 million km (932,000 mi) from Earth in the direction of the Sun. There, SOHO is able to observe the Sun continuously. Routine science operations began April 16, 1996. After December 21, 1998, all three gyros failed, but new software was uploaded; SOHO became the first three-axis spacecraft to operate without gyros.

SOHO has revolutionized our knowledge of the Sun. Major accomplishments include early warning of magnetic storms on Earth, detailed measurements beneath the Sun's surface, and images of the Sun's far side.

SOHO was built in Europe under overall management by ESA. The 12 instruments were provided by European and American scientists. NASA was responsible for the launch and is now responsible for mission operations. NASA's Deep Space Network is used for data downlink and commanding. Mission control is based at Goddard Space Flight Center in Maryland. The latest mission extension lasts until December 2009.

SPECIFICATION:

MANUFACTURER: Matra Marconi Space (now EADS Astrium)
LAUNCH DATE: December 2, 1995
ORBIT: L1 libration point halo
LAUNCH MASS: 1850 kg (4070 lb)
LAUNCH SITE: Cape Canaveral, Florida
LAUNCHER: Atlas II-AS
BODY DIMENSIONS: 4.3 x 2.7 x 3.7 m (14.1 x 8.9 x 12.1 ft)
PAYLOAD: Coronal Diagnostic Spectrometer (CDS); Charge, Element, and Isotope Analysis System (CELIAS); Comprehensive Suprathermal and Energetic Particle Analyzer (COSTEP); Extreme Ultraviolet Imaging Telescope (EIT); Energetic and Relativistic Nuclei and Electron experiment (ERNE); Global Oscillations at Low Frequencies (GOLF); Large Angle and Spectrometric Coronagraph (LASCO); Michelson Doppler Imager/Solar Oscillations Investigation (MDI/SOI); Solar Ultraviolet Measurements of Emitted Radiation (SUMER); Solar Wind Anisotropies (SWAN); Ultraviolet Coronagraph Spectrometer (UVCS); Variability of Solar Irradiance and Gravity Oscillations (VIRGO)

SORCE (Solar Radiation and Climate Experiment) USA

Solar radiation observatory

SORCE is part of NASA's Earth Observing System. Its five-year mission is to measure X-ray, ultraviolet, visible, near-infrared, and total solar radiation arriving at the Earth. The measurements address long-term climate change, natural variability and enhanced climate prediction, and atmospheric ozone and UV-B radiation. The hexagonal bus, which contains all control, propulsion, and communications sub-systems, is based on Orbital Sciences' LEOStar-2 platform. It is three-axis inertially stabilized. Precision pointing and attitude control are achieved by a combination of reaction wheels, two star trackers, a fine Sun sensor, a control processor, and torque rods and magnetometers. The S-band communications system uses dual transceivers and a pair of omnidirectional antennas.

SORCE was launched by a Pegasus XL dropped from an L-1011 aircraft over the Atlantic Ocean. An anomaly with its onboard computer caused a loss of data on May 15–20, 2007, but the spacecraft was fully recovered and returned to normal Sun-tracking mode.

SPECIFICATION:

MANUFACTURER: Orbital Sciences Corporation
LAUNCH DATE: January 25, 2003
ORBIT: 645 km (401 mi), 40° inclination
LAUNCH MASS: 290 kg (638 lb)
LAUNCH SITE: Cape Canaveral, Florida (air launch)
LAUNCHER: Pegasus XL
BODY DIMENSIONS: 1.6 x 1.2 m (5.2 x 3.8 ft)
PAYLOAD: Total Irradiance Monitor (TIM); Solar Stellar Irradiance Comparison Experiment (SOLSTICE); Spectral Irradiance Monitor (SIM); Extreme UV Photometer System (XPS)

Stardust USA

Comet flyby/comet dust/interplanetary dust sample return

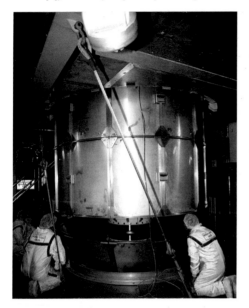

SPECIFICATION:

MANUFACTURER: Lockheed Martin Space Systems
LAUNCH DATE: February 7, 1999
ORBIT: Heliocentric
LAUNCH MASS: 385 kg (847 lb)
LAUNCH SITE: Cape Canaveral, Florida
LAUNCHER: Delta II 7926
BODY DIMENSIONS: 1.7 x 0.7 x 0.7 m (5.6 x 2.2 x 2.2 ft)
PAYLOAD: Aerogel dust collectors; Comet and Interstellar Dust Analyzer (CIDA); Navigation Camera (NavCam); Whipple Shield Dust Flux Monitors (DFM)

Stardust was a NASA Discovery mission. The spacecraft was derived from a rectangular bus called SpaceProbe, developed by Lockheed Martin. The sample-return capsule was a blunt-nosed cone with a diameter of 81 cm (32 in). It comprised a heat shield, back shell, sample canister, parachute system, and avionics.

Stardust had three dedicated science packages—a two-sided dust collector containing aerogel, a dust analyzer, and a dust flux monitor. Stardust completed three orbits of the Sun over seven years. On the way to Comet Wild 2, it traveled out to the asteroid belt and collected interstellar dust on two different solar orbits. In January 2000, its thrusters placed it on course for a gravity assist from Earth on January 15, 2001.

Stardust flew within 3100 km (1927 mi) of asteroid Annefrank on November 2, 2002. The comet flyby was on January 2, 2004, at a distance of 240 km (149 mi). Stardust then returned to Earth. On January 14, 2006, the sample-return capsule parachuted safely onto the salt flats of Utah after the fastest-ever spacecraft reentry (46,444 km/h or 28,860 mph). The main spacecraft maneuvered into heliocentric orbit and will take part in the Stardust-NExT follow-on mission to fly past Comet Tempel 1 on February 14, 2011.

STEREO (Solar Terrestrial Relations Observatory) USA

Dual solar observatory

SPECIFICATION:

MANUFACTURER: Johns Hopkins University Applied Physics Laboratory
LAUNCH DATE: October 26, 2006
ORBIT: Heliocentric, 149.6 million km (93 million mi)
LAUNCH MASS: 620 kg (1364 lb) (x2)
LAUNCH SITE: Cape Canaveral, Florida
LAUNCHER: Delta II 7925-10L
BODY DIMENSIONS: 1.1 x 1.2 x 2 m (3.7 x 4 x 6.6 ft)
PAYLOAD: Sun-Earth Connection Coronal and Heliospheric Investigation (SECCHI)—extreme ultraviolet imager; 2 white-light coronagraphs; heliospheric imager; In Situ Measurements of Particles and CME Transients (IMPACT); Plasma and Suprathermal Ion Composition (PLASTIC); STEREO/WAVES radio burst tracker (S/WAVES)

STEREO is the third mission in NASA's Solar Terrestrial Probes Program and is the first mission to produce three-dimensional images of the Sun.

The two observatories each have two solar panels, a high-gain dish antenna, three wire antennas, and an instrument boom. Each carries four science packages made up of several instruments to study the evolution and motion of huge eruptions of solar plasma called coronal mass ejections, including their effects on Earth's magnetic field.

After launch, both spacecraft completed a series of four highly elliptical phasing orbits around Earth to position them for a lunar gravitational assist. On December 15, 2006, STEREO "Ahead" flew within 7340 km (4550 mi) of the Moon and used lunar gravity to move into its orbit ahead of Earth. STEREO "Behind" obtained a smaller gravity assist from the Moon on December 15, followed by a 8818 km (5468 mi) flyby on January 21, 2007, which sent it into an orbit behind Earth. The observatories separate by 45° every year and will eventually be behind the Sun. The first three-dimensional solar images were released on April 23, 2007. Mission duration is two years.

THEMIS (Time History of Events and Macroscale Interactions During Substorms) USA

Auroral studies

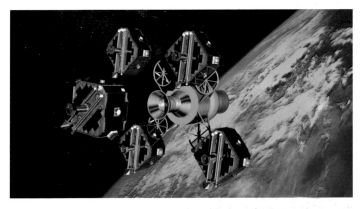

NASA's first five-satellite mission investigates how Earth's auroras are triggered by the storage and release of energy in the magnetosphere. Data collected from THEMIS helps pinpoint where and when substorms begin.

Each of the identical small spacecraft is spin-stabilized. The cube-shaped bus is composed of a graphite structure and has four side panels with face-mounted solar arrays. Each carries 11 instruments.

The spacecraft were deployed into orbit from a rotating platform on the launcher's upper stage. Upon reaching their operational orbits in late 2007, the constellation will line up at different distances along the Sun–Earth line. Their data will be combined with measurements from imaging systems across Alaska and Canada to document how the northern aurora appears.

The launch of THEMIS marked the first collaboration between NASA and the United Launch Alliance of Lockheed Martin and Boeing. The mission control center is at the University of California, Berkeley.

SPECIFICATION:

MANUFACTURER: Swales Aerospace
LAUNCH DATE: February 17, 2007
ORBIT: Probe 1, 9567 x 201,831 km (5945 x 125,412 mi), 7° inclination; Probe 2, 7450 x 126,093 km, 7° inclination; Probe 3, 7130 x 77,346 km (4629 x 78,350 mi), 9° inclination; Probe 4, 7130 x 77,346 km (4430 x 48,060 mi), 9° inclination; Probe 5, 8610 x 64,060 km (5350 x 39,805 mi), 4.5° inclination
LAUNCH MASS: 128 kg (282 lb) (x 5)
LAUNCH SITE: Cape Canaveral, Florida
LAUNCHER: Delta II 7925
BODY DIMENSIONS: 0.8 x 0. x 0.5 m (2.8 x 2.8 x 1.7 ft)
PAYLOAD: Instrument Data Processing Unit (IDPU); 6 Electric Field Instruments (EFI); Fluxgate Magnetometer (FGM); Search Coil Magnetometer (SCM); Electrostatic Analyzer (ESA); 2 Solid State Telescopes (SST)

TRACE (Transition Region and Coronal Explorer) USA

Solar observatory

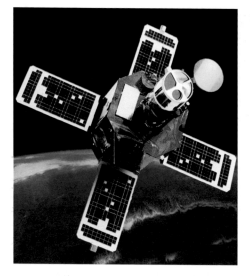

SPECIFICATION:

MANUFACTURER: Lockheed Martin
LAUNCH DATE: April 2, 1998
ORBIT: 600 x 650 km (373 x 404 mi), 97.8° inclination (Sun-synchronous)
LAUNCH MASS: 250 kg (550 lb)
LAUNCH SITE: Vandenberg, California
LAUNCHER: Pegasus XL
BODY DIMENSIONS: 1.8 x 1.1 m (5.9 x 3.5 ft)
PAYLOAD: Extreme-Ultraviolet Imaging Telescope (EIT)

TRACE is a NASA Small Explorer mission to image the solar corona and transition region at high angular and temporal resolution. The mission is complementary to SOHO. It was launched on a Pegasus XL rocket released over the Pacific from an L-1011 aircraft. Its polar orbit enables virtually continuous observations of the Sun, uninterrupted by the Earth's shadow for long periods. This orbit gives the greatest chance of observing the processes that lead to flares and massive eruptions in the Sun's atmosphere.

TRACE is a single-instrument, three-axis-stabilized spacecraft. The spacecraft attitude control system uses three magnetic-torquer coils, a digital Sun sensor, six coarse Sun sensors, a three-axis magnetometer, four reaction wheels, and three gyros to maintain pointing.

The octagonal spacecraft bus contains the electronics, computers, transponder, gyros, reaction wheels, and torque rods. Attached to it are four gallium arsenide solar arrays and two antennas (located at the end of two arrays). On top is the main science instrument, a 30 cm (12 in) telescope that images the Sun in ultraviolet light. The telescope was designed, fabricated, tested, and calibrated by the Solar and Astrophysics Laboratory at Lockheed Martin's Advanced Technology Center in collaboration with Stamford University. The baseline mission was one year, but the satellite continued in good health after nine years.

Ulysses EUROPE/USA

First solar polar orbiter

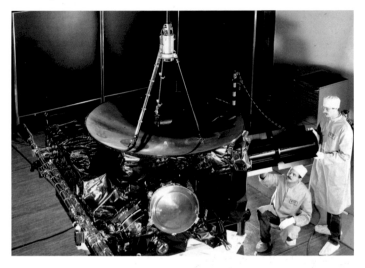

Ulysses was originally part of an ESA (with NASA) two-spacecraft International Solar Polar Mission. After the NASA mission was canceled, the US collaborated on ESA's Ulysses. Ulysses was the fastest spacecraft to leave Earth and the only spacecraft to fly over high solar latitudes, as far as 80° N and S. A gravity assist from Jupiter on February 8, 1992, directed it away from the ecliptic into a 6.2-year orbit around the Sun. It has now observed solar activity during one complete solar cycle (11 years). In 1996, Ulysses unexpectedly crossed Comet Hyakutake's tail—the longest tail ever recorded. Key mission results include the first detailed measurements of the solar wind from the Sun's polar regions at solar minimum and solar maximum, the discovery that the magnetic flux leaving the Sun is the same at all latitudes, and the first direct measurements of interstellar helium atoms in the solar system. Ulysses is now on its third orbit over the Sun's poles. It is currently funded until March 2009.

SPECIFICATION:

MANUFACTURER: Dornier
LAUNCH DATE: October 6, 1990
ORBIT: Heliocentric, 200 million x 808 million km (124.6 million x 502 million mi), 79.1° inclination
LAUNCH MASS: 370 kg (814 lb)
LAUNCH SITE: Kennedy Space Center, Florida
LAUNCHER: Space Shuttle *Discovery* (STS-41)
BODY DIMENSIONS: 3.2 x 3.3 x 2.1 m (10.5 x 10.8 x 6.9 ft)
PAYLOAD: Vector Helium Magnetometer/Fluxgate Magnetometer (VHM/FGM); Solar Wind Observations Over the Poles of the Sun (SWOOPS); Solar Wind Ion Composition Spectrometer (SWICS); Unified Radio and Plasma Wave Investigation (URAP); Energetic Particles Composition/Inter-stellar Neutral Gas EPAC/GAS; Heliosphere Instrument for Spectra, Composition and Anisotropy at Low Energies (HI-SCALE); Cosmic Ray and Solar Particle Investigation (COSPIN); Gamma-Ray Burst experiment (GRB); Cosmic Dust (DUST)

Venus Express EUROPE

Venus orbiter

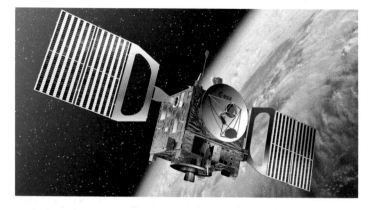

Since Venus Express spacecraft is a virtual twin of Mars Express, it was developed very quickly and at low cost. The major differences are in thermal control with larger, more efficient radiators and 23 layers of multilayer insulation that are gold instead of black. The two lateral solar arrays were also redesigned.

The scientific instruments are located on three sides of the spacecraft. There are two high-gain antennas—a 1.3 m (4.3 ft) dish and a 0.3 m (1 ft) antenna—mounted on different faces.

Venus Express arrived at Venus on April 11, 2006, after a five-month journey. The main engine was fired twice (on April 20 and 23) and the thrusters were ignited five times (April 15, 26, and 30, and May 3 and 6). On May 7, 2006, after 16 loops around the planet, Venus Express entered its nominal science operations orbit. Pericenter is located at 80° N and each orbit lasts 24 hours. Joint observations were undertaken with NASA's Messenger during its Venus flyby on June 6, 2007. The mission is expected to last for at least two Venus sidereal days (486 Earth days).

SPECIFICATION:

MANUFACTURER: EADS Astrium
LAUNCH DATE: November 9, 2005
OPERATIONAL VENUS ORBIT: 250 x
66,000 km (155 x 41,010 mi),
90° inclination
LAUNCH MASS: 1270 kg (2794 lb)
LAUNCH SITE: Baikonur,
Kazakhstan
LAUNCHER: Soyuz Fregat
BODY DIMENSIONS: 1.5 x 1.8 x
1.4 m (4.9 x 5.9 x 4.6 ft)
PAYLOAD: Analyzer of Space
Plasma and Energetic Atoms
(ASPERA); Venus Express
Magnetometer (MAG); Planetary
Fourier Spectrometer (PFS);
Ultraviolet and Infrared
Atmospheric Spectrometer
(SPICAV/SOIR); Venus Radio
Science Experiment (VeRa);
Ultraviolet/Visible/Near-Infrared
Mapping Spectrometer (VIRTIS);
Venus Monitoring Camera
(VMC)

Wind USA

Solar wind observer

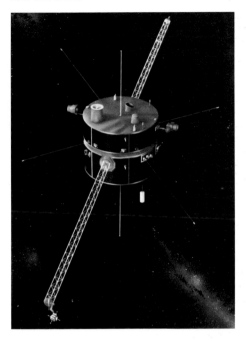

SPECIFICATION:

MANUFACTURER: Martin Marietta
(now Lockheed Martin)
LAUNCH DATE: November 1, 1994
INITIAL ORBIT: 48,840 x
1,578,658 km (30,340 x
980,930 mi), 19.65° inclination
(variable)
LAUNCH MASS: 1195 kg (2635 lb)
LAUNCH SITE: Cape Canaveral,
Florida
LAUNCHER: Delta II 7925
BODY DIMENSIONS: 2.4 x 1.8 m
(7.9 x 5.9 ft)
PAYLOAD: Magnetic Field
Investigation (MFI); Radio and
Plasma Wave Experiment
(WAVES); Solar Wind Experiment
(SWE); Energetic Particle
Acceleration, Composition, and
Transport (EPACT); Solar Wind
and Suprathermal Ion Compo-
sition Studies (SWICS/STICS);
3-Dimensional Plasma and
Energetic Particle Analyzer (3-D
Plasma); Transient Gamma-Ray
Spectrometer (TGRS); Gamma-
Ray Burst Detector (KONUS)

Wind studies the solar wind, both inside and outside Earth's magnetosphere. With its sister satellite Polar, it is a key component of the International Solar-Terrestrial Physics program. Wind's main scientific goal was to measure the mass, momentum, and energy of the solar wind.

Later in its mission, Wind was inserted into a halo orbit in the solar wind upstream from the Earth, about the sunward Sun–Earth equilibrium point (L1 Lagrange point). After several months at L1, Wind made two passes of the Moon to begin a series of "petal" orbits that took it up to 60° out of the ecliptic plane. On August 19, 2000, it flew within 7600 km (4723 mi) of the Moon and subsequently moved to an approximately 567,000 x 1,620,000 km (352,000 x 1,006,650 mi), 21.8° orbit.

The drum-shaped satellite has a coating of solar cells. It is spin-stabilized with a spin period of about 20 sec. It has two 12 m (39.4 ft) radial booms for magnetometer and waves sensors. The wire antennae are 50 m (164 ft) and 7.5 m (24.6 ft) in length. Wind carried the first Russian instrument (KONUS) to fly on a US spacecraft after cooperation resumed in 1987.

Earth
Remote
Sensing
Satellites

AIM (Aeronomy of Ice in the Mesosphere) USA

Upper-atmosphere studies

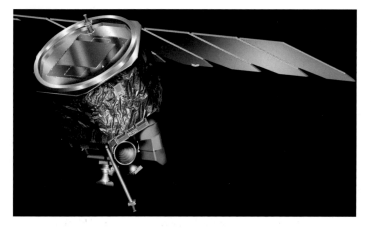

NASA's AIM spacecraft carries three instruments to study polar mesospheric clouds located 80 km (50 mi) above the Earth's surface in the coldest part of the atmosphere. Its primary goal is to explain why these clouds form and what has caused them to become brighter and more numerous and to appear at lower latitudes in recent years.

AIM was air-launched by a Pegasus XL carried aloft over the Pacific Ocean by a modified L-1011 aircraft. The satellite was released into a circular near-polar orbit that gives it the best view of the night-shining clouds. AIM will spend two years studying the clouds.

AIM carries three instruments. The Cloud Imaging and Particle Size camera system obtains panoramic images of the clouds. The Solar Occultation for Ice Experiment measures the reduction in the Sun's light to determine the size and density of particles in the clouds, along with the temperature, water-vapor levels, and chemistry. The Cosmic Dust Experiment measures the amount of space dust entering the atmosphere. AIM is the ninth NASA Small Explorer mission.

SPECIFICATION:

MANUFACTURER: Orbital Sciences Corporation
LAUNCH DATE: April 25, 2007
ORBIT: 600 km (373 mi), 97.7° inclination (Sun-synchronous)
LAUNCH MASS: 195 kg (430 lb)
LAUNCH SITE: Vandenberg, California (air launch)
LAUNCHER: Pegasus XL
BODY DIMENSIONS: 1.4 x 1.1 m (4.6 x 3.6 ft)
PAYLOAD: Cloud Imaging and Particle Size Experiment (CIPS); Solar Occultation for Ice Experiment (SOFIE); Cosmic Dust Experiment

ALOS/Daichi (Advanced Land-Observing Satellite) JAPAN

Earth remote sensing

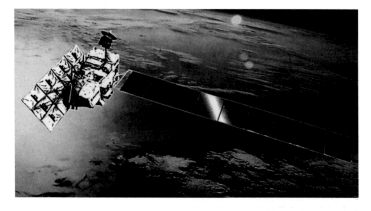

The Advanced Land-Observing Satellite, also known as Daichi (Land), is one of the largest Earth-observing satellites ever built. It was developed for normal and three-dimensional surface mapping, precise regional land coverage observation, disaster monitoring, and resource surveying. Its design is influenced by experience gained from its predecessors, the Japanese Earth Resource Satellite-1 (JERS-1/Fuyo) and the two Advanced Earth-Observing Satellites (ADEOS/Midori). ALOS has three sensors: a 2.5 m (8.25 ft) resolution stereo mapping camera to measure precise land elevation, a 10 m (32.8 ft) resolution visible and near-infrared radiometer that observes land cover, and a synthetic aperture radar that enables day–night and all-weather land observation.

On April 5, 2007, the observation data relay from ALOS to the Data Relay Satellite Kodama was interrupted. The ALOS data relay transmission system was subsequently switched to the back-up system to restore normal operations.

SPECIFICATION:

MANUFACTURER: JAXA
LAUNCH DATE: January 24, 2006
ORBIT: 692 km (430 mi), 98.2° (Sun-synchronous)
LAUNCH MASS: 3850 kg (8470 lb)
LAUNCH SITE: Tanegashima, Japan
LAUNCHER: H-IIA-8
BODY DIMENSIONS: Approx. 4 x 4 x 6 m (13.1 x 13.1 x 19.7 ft)
PAYLOAD: Panchromatic Remote-sensing Instrument for Stereo Mapping (PRISM); Advanced Visible and Near-Infrared Radiometer Type 2 (AVNIR-2); Phased Array type L-band Synthetic Aperture Radar (PALSAR)

AlSAT ALGERIA

Disaster monitoring

AlSAT-1 is Algeria's first national satellite and the first satellite in the international Disaster Monitoring Constellation (DMC). It was designed and constructed by SSTL at the Surrey Space Center (UK) as part of a joint program with the Algerian Centre National des Techniques Spatiales (CNTS). A British–Algerian team of SSTL and CNTS engineers completed the manufacture and pre-flight testing of the satellite. A control center was also installed in Algeria.

AlSAT-1 is based on the SSTL Microsat-100 bus, but carries a "push broom" imager—a first for SSTL microsatellites. The imaging system comprises two cameras for each spectral band and two sensors with 10,000 pixels each. These give 32 m (105 ft) spatial resolution imaging in three spectral bands (near-infrared, red, and green) with an imaging swath of 600 km (373 mi) on the ground. This enables a revisit of the same area anywhere in the world at least every four days. Three solid-state data recorders allow one GB storage for images. As part of the DMC constellation, AlSAT-1 has a thruster system for small orbit corrections. Two more AlSATs are on order with EADS Astrium.

SPECIFICATION:

MANUFACTURER: Surrey Satellite Technology Limited
LAUNCH DATE: November 28, 2002
ORBIT: 686 km (426 mi), 98.2° inclination (Sun-synchronous)
LAUNCH SITE: Plesetsk, Russia
LAUNCHER: Kosmos-3M
LAUNCH MASS: 90 kg (198 lb)
BODY DIMENSIONS: 0.6 x 0.6 x 0.6 m (2 x 2 x 2 ft)
PAYLOAD: Multi-spectral CCD imager

Aqua USA

Global water monitoring

Aqua (Latin for "water") is a joint project involving the US, Japan, and Brazil. It is the second major platform in NASA's series of Earth-Observing System spacecraft and part of NASA's Earth Science Enterprise. It is the sister to NASA's Terra spacecraft, launched in December 1999. The US provided the spacecraft and four of the six scientific instruments. AMSR-E was built by Mitsubishi Electronics Corp. and provided by the Japanese National Space Development Agency. The HSB was built by Matra Marconi Space (now EADS Astrium) and provided by the Brazilian Space Agency.

Aqua is based on TRW's modular, standardized T-330/AB1200 spacecraft bus. The spacecraft was built of lightweight composite materials and is designed for a six-year mission. Formerly named EOS/PM-1, Aqua crosses the equator daily at 1.30 pm as it heads north. Flying in a near-polar, Sun-synchronous orbit means that its afternoon observations, combined with Terra's morning observations, provide important data about daily changes in global precipitation, evaporation, and the cycling of water in the atmosphere, cryosphere, on land, and in the oceans. It returns about 89 GB of data daily.

SPECIFICATION:

MANUFACTURER: TRW (now Northrop Grumman)
LAUNCH DATE: May 4, 2002
ORBIT: 699 x 706 km (434 x 438 mi), 98° inclination (Sun-synchronous)
LAUNCH SITE: Vandenberg, California
LAUNCHER: Delta II-7920
LAUNCH MASS: 2934 kg (6468 lb)
BODY DIMENSIONS: 2.7 x 2.5 x 6.5 m (8.8 x 8.2 x 21.3 ft)
PAYLOAD: Atmospheric Infrared Sounder (AIS); Moderate-Resolution Imaging Spectro-radiometer (MODIS); Advanced Microwave Sounding Unit (AMSU); Clouds and the Earth's Radiant Energy System (CERIS); Advanced Microwave Scanning Radiometer for Earth-Observing Satellite (AMSR-E); Humidity Sounder for Brazil (HSB)

Aquarius (SAC-D) USA/ARGENTINA

Measurement of global sea surface salinity

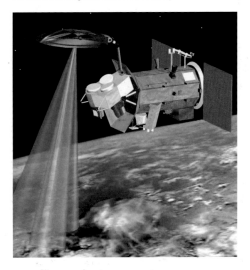

SPECIFICATION:

MANUFACTURER: CONAE
(Argentina)
LAUNCH DATE: 2009?
ORBIT: 657 km (408 mi), 98°
inclination (Sun-synchronous)
LAUNCH MASS: 1642 kg (3612 lb)
LAUNCH SITE: Vandenberg,
California
LAUNCHER: Delta II-7320
BODY DIMENSIONS: 2.4 x 4.9 m
(9 x 15.9 ft)
PAYLOAD: Microwave Radiometer
(MWR); New Infrared Scanner
Technology radiometer (NIRST);
Aquarius L-band radiometer/
scatterometer; High-Sensitivity
Camera (HSC); Radio
Occultation Sounder of the
Atmosphere (ROSA); CARMEN-1
radiation monitoring—orbital
debris/micrometeorite detector;
orbit determination Technology
Demonstration Package (TDP)

Aquarius is a joint NASA–Space Agency of Argentina (Comisión Nacional de Actividades Espaciales [CONAE]) mission to measure global sea surface salinity (SSS). It is known as Aquarius in the US, after its primary instrument, and as SAC-D (Scientific Application Satellite-D) in Argentina. It is one of two new Earth System Science Pathfinder small-satellite missions confirmed by NASA in 2005.

The primary objective is to study the processes related to the water cycle and ocean circulation by measuring global variations in sea surface salinity for at least three years. It will study how the ocean responds to the combined effects of evaporation, precipitation, ice melt, and river runoff. The Aquarius instrument will provide maps of salt concentration on the ocean surface. The near-polar orbit will provide complete Earth coverage every eight days.

The spacecraft is three-axis stabilized, using thrusters for orbit maintenance. CONAE will provide the spacecraft, based on the SAC-C platform, and five payload instruments. It will also handle ground operations, including receipt of telemetry and scientific data. One payload each will be provided by ASI (Italy) and CNES (France).

Aura USA

Study of atmospheric chemistry

Aura (formerly EOS/Chem-1) was the third major mission in NASA's Earth-Observing System, following on from Terra and Aqua. Its purpose is to study the chemistry and dynamics of Earth's atmosphere from the ground to the mesosphere. The goal is to monitor the complex chemical interactions of atmospheric constituents produced by natural and manmade sources, and their contributions to global change and depletion of the ozone layer.

The lightweight spacecraft structure is made of graphite epoxy composite over a honeycomb core. A solar array provides 4600 W of power. The onboard data system can store over 100 Gb of data. The spacecraft crosses the equator at 1.45 pm, plus or minus 15 min, repeating its ground track every 16 days. This enables it to make repeat atmospheric measurements over virtually the entire Earth under similar lighting conditions. Aura's limb instruments are all designed to observe roughly along the orbit plane. MLS is on the front of the spacecraft (the forward velocity direction), while HIRDLS, TES, and OMI are mounted on the nadir (Earth-facing) side. The satellite was designed to operate for at least six years.

SPECIFICATION:

MANUFACTURER: TRW (now Northrop Grumman)
LAUNCH DATE: July 15, 2004
ORBIT: 688 x 694 km (427 x 431 mi), 98.2° inclination (Sun-synchronous)
LAUNCH SITE: Vandenberg, California
LAUNCHER: Delta II-7920
LAUNCH MASS: 2967 kg (6542 lb)
BODY DIMENSIONS: 2.7 x 2.3 x 6.9 m (8.8 x 7.5 x 22.7 ft)
PAYLOAD: High Resolution Dynamics Limb Sounder (HIRDLS); Microwave Limb Sounder (MLS); Ozone Monitoring Instrument (OMI); Tropospheric Emission Spectrometer (TES)

CALIPSO (Cloud-Aerosol Lidar and Infrared Pathfinder Satellite Observation) USA/FRANCE

Measurement of aerosols and clouds

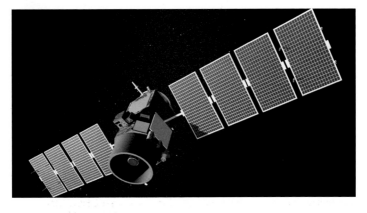

CALIPSO is a joint mission involving NASA and CNES, the French space agency. Its purpose is to provide global measurements of aerosols and clouds in order to improve predictions of long-term climate change and climate variability. CNES provided a PROTEUS spacecraft bus, developed by Alcatel Alenia Space, and the radiometer instrument. It also handles mission operations. The payload consists of three coaligned nadir-viewing instruments. Science data are downlinked using an X-band transmitter. Ball Aerospace developed the lidar (light detection and ranging) instrument and visible camera. Together, the instruments measure vertical distributions of aerosols and clouds in the atmosphere, as well as the optical and physical properties of aerosols and clouds.

CALIPSO was launched together with NASA's CloudSat. Both are members of NASA's A-Train constellation. It is on a three-year mission and repeats the same ground track every 16 days.

SPECIFICATION:

MANUFACTURER: Alcatel Alenia Space
LAUNCH DATE: April 28, 2006
ORBIT: 702 km (436 mi), 98.2° inclination (Sun-synchronous)
LAUNCH SITE: Vandenberg, California
LAUNCHER: Delta II-7420
LAUNCH MASS: 587 kg (1294 lb)
BODY DIMENSIONS: 1.9 x 1.6 x 2.5 m (6.3 x 5.3 x 8.1 ft)
PAYLOAD: Cloud-Aerosol Lidar with Orthogonal Polarization (CALIOP); Wide-Field Camera (WFC); Imaging Infrared Radiometer (IIR)

CARTOSAT INDIA

Earth observation and mapping

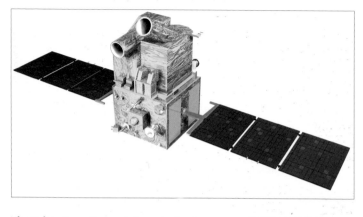

The Indian Space Research Organization has launched two CARTOSAT spacecraft for Earth observation. CARTOSAT-1 (IRS-P5), which weighed about 1560 kg (3432 lb) at launch, is a stereoscopic Earth-observation satellite in a Sun-synchronous polar orbit. Mainly used for cartography and terrain modeling, it was launched by the PSLV on May 5, 2005. It carries two panchromatic cameras that that cover a 30 km (18.6 mi) wide swath with a spatial resolution of 2.5 m (8.2 ft). The cameras can be steered across the direction of the satellite's movement and are mounted so that near-simultaneous imaging of the same area from two different angles is possible.

CARTOSAT-2 is the twelfth in the Indian Remote Sensing (IRS) satellite series. It also carries a panchromatic camera with a spatial resolution of better than 1 m (3.3 ft) and a swath width of 9.6 km (6 mi). The satellite can be steered up to 45° along as well as across the track. It revisits the same location every four days. The data are used for detailed mapping and other cartographic applications. Design life is five years.

SPECIFICATION (CARTOSAT-2):

MANUFACTURER: ISRO
LAUNCH DATE: January 10, 2007
ORBIT: 635 km (395 mi), 97.9° inclination (Sun-synchronous)
LAUNCH MASS: 680 kg (1496 lb)
LAUNCH SITE: Sriharikota, India
LAUNCHER: PSLV-C7
PAYLOAD: Panchromatic camera

CBERS (China–Brazil Earth Resources Satellite/ZiYuan)

BRAZIL/PEOPLE'S REPUBLIC OF CHINA

Earth remote sensing

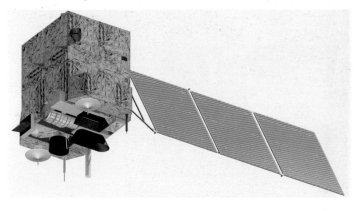

Three CBERS (known in China as ZiYuan, or "Earth Resources") have been jointly developed by Brazil and China to monitor natural resources and the environment. The first (CBERS-1/ZiYuan-1) was launched on October 14, 1999, by a Long March 4B from Taiyuan. The satellite was composed of two modules. The payload module carried three cameras for imaging at visible and infrared wavelengths, and a transponder for the Brazilian environmental data collection system. China built the satellite and the primary imaging system; Brazil provided the wide-field imager and main satellite subsystems. From a Sun-synchronous orbit, it obtained complete coverage of the Earth in 26 days. Image swath was 120 km (75 mi) across with the high-resolution camera, and best spatial resolution was 19.5 m (64 ft).

CBERS-2/ZiYuan-1B was launched on October 21, 2003. It is technically identical to CBERS-1: two solar panels provide 1.1 kW of power; hydrazine thrusters are used for attitude control and orbit maneuvers. Nominal design life is two years. CBERS-2B was launched from Taiyuan on September 19, 2007.

SPECIFICATION (CBERS-1):

MANUFACTURERS: ADE, Funcate, Digicon, CAST
LAUNCH DATE: October 14, 1999
ORBIT: 778 km (483 mi), 98.5° inclination (Sun-synchronous)
LAUNCH MASS: 1450 kg (3190 lb)
LAUNCH SITE: Taiyuan, China
LAUNCHER: Long March 4B
BODY DIMENSIONS: 1.8 x 2 x 2.2 m (5.9 x 6.6 x 7.2 ft)
PAYLOAD: High-resolution CCD cameras; Infrared Multispectral Scanner (IRMSS); Wide-Field Imager (WFI)

CloudSat USA

Three-dimensional cloud studies

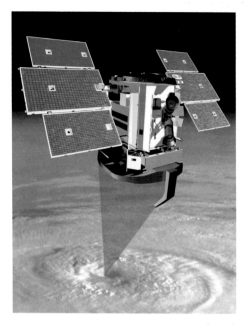

SPECIFICATION:

MANUFACTURER: Ball Aerospace
LAUNCH DATE: April 28, 2006
ORBIT: 705 km (438 mi),
98.2° inclination
(Sun-synchronous)
LAUNCH MASS: 848 kg (1870 lb)
LAUNCH SITE: Vandenberg,
California
LAUNCHER: Delta II-7420
BODY DIMENSIONS: 2.5 x 2 x
2.3 m (8.3 x 6.7 x 7.5 ft)
PAYLOAD: Cloud-Profiling Radar
(CPR)

NASA's CloudSat provides the first global survey of cloud profiles and properties, using advanced radar to see their vertical structure. CloudSat was launched together with the NASA–CNES CALIPSO satellite. Both satellites are members of NASA's A-Train constellation of five Earth-Observing System satellites. CloudSat flies in tight formation with CALIPSO, and both satellites follow behind Aqua. The same ground track is repeated every 16 days.

The satellite is based on the Ball Aerospace RS2000 bus, used previously for both NASA's QuikSCAT and ICESat missions. Two solar arrays provide 570 W of power. Attitude control and maneuvering is carried out by hydrazine thrusters. CloudSat is managed by NASA's Jet Propulsion Laboratory, which also developed the 94 GHz cloud-profiling radar instrument. Other hardware contributions come from the Canadian Space Agency. Colorado State University, the US Air Force, and the US Department of Energy have also contributed. Design life is 22 months.

Coriolis USA

Global wind measurement and geomagnetic storm warning

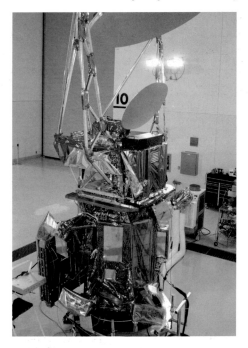

SPECIFICATION:

MANUFACTURER: Spectrum Astro (now General Dynamics C4 Systems)
LAUNCH DATE: January 6, 2003
ORBIT: 830 km (438 mi), 98.2° inclination (Sun-synchronous)
LAUNCH SITE: Vandenberg, California
LAUNCHER: Titan 2
LAUNCH MASS: 827 kg (1820 lb)
BODY DIMENSIONS: 4.7 x 1.3 m (15.4 x 4.4 ft)
PAYLOAD: WindSat microwave polarimetric radiometer; Solar Mass Ejection Imager

Coriolis was developed by the US DOD Space Test Program and the Space and Naval Warfare Systems Command. It was the largest spacecraft ever launched by a refurbished Titan 2 missile. The size of the spacecraft is determined by the WindSat instrument, which is about 3.3 m (11 ft) tall, with a truss-mounted 2 m (6 ft) dish antenna that spins at 31.6 rpm to provide maximum Earth coverage.

Its mission is to demonstrate remote sensing of global wind vectors using the Microwave Polarimetric Radiometry technique, as well as risk reduction for the Conical Microwave Imager Sounder to be flown on the future National Polar-orbiting Operational Environmental Satellite System (NPOESS). It will also continuously observe solar mass ejections in order to more rapidly and accurately predict geomagnetic disturbances that can affect satellites. The US Navy's WindSat instrument measures wind speed and direction at or near the ocean surface. The US Air Force Solar Mass Ejection Imager provides early warning of coronal mass ejections. Two solar arrays provide 1.17 kW of power. Design life is three years.

CryoSat EUROPE

Measurement of ice thickness and cover

CryoSat was intended to be the first Earth Explorer mission in ESA's Living Planet program. As the result of a Rockot launcher malfunction, the first CryoSat plunged into the Arctic Ocean in October 2005. A few months later, ESA contracted with Astrium to rebuild the satellite. The design of CryoSat-2 is based on CryoSat-1, although there have been some 85 modifications. One of the main changes is the inclusion of a back-up for the main payload, the SAR/Interferometric Radar Altimeter, with a second set of electronics.

CryoSat has no moving parts at all, except for some valves in the propulsion system. It has two solar panels fixed to the satellite body, forming an angled "roof." These provide 1.6 kW of power. An S-band helix antenna on the satellite's underside is used for systems monitoring and commands from the ground. Next to it is an X-band antenna that transmits data to the ground station at Kiruna, Sweden. The radar instrument will monitor the thickness of ice cover on land and sea in order to refine the link between melting polar ice and the rise in sea levels and its possible contribution to climate change. The mission lifetime is three years.

SPECIFICATION (CryoSat-1):

MANUFACTURER: EADS Astrium
LAUNCH DATE: October 8, 2005
INTENDED ORBIT: 717 km (446 mi), 92° inclination
LAUNCH MASS: 684 kg (1505 lb)
LAUNCH SITE: Plesetsk, Russia
LAUNCHER: Rockot
BODY DIMENSIONS: 4.6 x 2.3 x 2.2 m (15.1 x 7.7 x 7.2 ft)
PAYLOAD: 2 SAR/Interferometric Radar Altimeters (SIRAL); 7 laser retroflectors (LRR); Doppler Orbit and Radio Positioning (DORIS)

Disaster Monitoring Constellation (UK-DMC) UK

Disaster monitoring

SPECIFICATION (UK-DMC):

MANUFACTURER: Surrey Satellite Technology Limited
LAUNCH DATE: September 27, 2003
ORBIT: 675 x 692 km (419 x 430 mi), 98.2° (Sun-synchronous)
LAUNCH SITE: Plesetsk, Russia
LAUNCHER: Kosmos 3M
LAUNCH MASS: 80 kg (176 lb)
BODY DIMENSIONS: 0.6 x 0.6 x 0.6 m (2 x 2 x 2 ft)
PAYLOAD: Multispectral CCD imager; Internet router

After the launch of AlSAT-1, three more Disaster Monitoring Constellation satellites were launched in September 2003 into four equispaced slots in one orbital plane. They were UK-DMC (funded by the UK government), BILSAT-1 for Turkey (funded by the Tubitak-ODTU Bilten), and NigeriaSat-1 for Nigeria (funded by the National Space Research and Development Agency).

UK-DMC uses the MicroSat-100 bus and was developed for the British National Space Center under a grant from the Microsatellite Applications in Collaboration (MOSAIC) program. Like all of the DMC satellites, its optical imaging payload provides 32 m (105 ft) ground resolution with an exceptionally wide swath of over 640 km (398 mi). The payload uses green, red, and near-infrared bands equivalent to Landsat TM+ bands 2, 3, and 4.

Compared with the other DMC satellites, UK-DMC has increased onboard data storage, with 1.5 GB capacity. Images are returned to the operations center using Internet protocol over an 8 Mbps S-band downlink. UK-DMC also has a commercial router to experiment with Internet packet routing to and in space. In addition, it provided an opportunity to test GPS reflectometry, a new concept in oceanographic remote sensing, which measures GPS signals from the navigation system after they are reflected off the sea.

Envisat EUROPE

Earth remote sensing

Envisat is the largest and most complex satellite ever built by ESA. It is based on the Polar Platform originally developed for the Columbus program. The platform, based on the Spot 4 service module, provides power and communications. A single solar array generates 6.5 kW. On the nadir side are a 10 m (33 ft) long ASAR antenna and most of the science instruments. The satellite flies on a 35-day repeat cycle, 30 min ahead of ERS-2, with a 10.00 am descending node.

The 10 instruments address four main areas: radar imaging; optical imaging over oceans, coastal zones, and land; observation of the atmosphere; and altimetry. About two-thirds of the data are transmitted to the ground via ESA's Artemis relay satellite, providing data acquisition for any location worldwide. Typical data download rate is 250 GB per day. Envisat data are used in many fields of Earth science, including atmospheric pollution, fire extent, sea-ice motion, ocean currents, and vegetation change, as well as for operational activities such as mapping land subsidence, monitoring oil slicks, and watching for illegal fisheries. Nominal mission lifetime was five years.

SPECIFICATION:

MANUFACTURER: Dornier, Matra Marconi Space (now EADS Astrium)
LAUNCH DATE: March 1, 2002
ORBIT: 784 km (487 mi) 98.5° inclination (Sun-synchronous)
LAUNCH MASS: 8111 kg (17,844 lb)
LAUNCH SITE: Kourou, French Guiana
LAUNCHER: Ariane 5G
BODY DIMENSIONS: 10.5 x 4.6 m (34.4 x 14.7 ft)
PAYLOAD: Advanced Synthetic Aperture Radar (ASAR); Advanced Along-Track Scanning Radiometer (AATSR); Medium-Resolution Imaging Spectrometer (MERIS); Advanced Radar Altimeter (RA-2); Microwave Radiometer (MWR); Michelson Interferometer for Passive Atmospheric Sounding (MIPAS); Scanning Imaging Absorption Spectrometer for Atmospheric Cartography (SCIAMACHY); Global Ozone Monitoring by Occultation of Stars (GOMOS); Doppler Orbit and Radio Positioning (DORIS); Laser retroflectors (LRR)

EROS (Earth Remote Observation Satellite) ISRAEL

Earth remote sensing

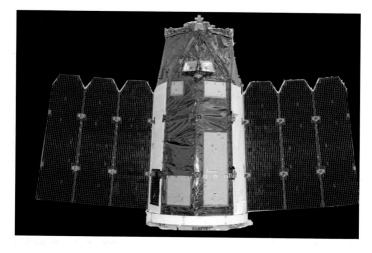

The two EROS satellites are high-resolution commercial imaging satellites, derived from the Israeli Ofeq military reconnaissance satellites. Both use the same bus and are structurally identical. They are owned and operated by ImageSat International of Israel, which was the first non-US company to successfully deploy a commercial high-resolution imaging satellite. The satellites are mainly marketed for military use (views of Israel are excluded).

EROS A was launched in December 2000 by a Russian Start-1, a rare commercial launch from Svobodny. The spacecraft is three-axis stabilized with two lateral solar arrays. Its lightweight design means that it can be quickly pointed to customer-specified sites at angles of up to 45°. Oblique viewing enables it to view virtually any site two to three times a week. EROS B has a larger CCD camera with panchromatic resolution of 0.7 m (2.3 ft), a larger onboard recorder, and a faster data communication link. Design life is 8–10 years.

SPECIFICATION (EROS B):

MANUFACTURER: Israel Aircraft Industries
LAUNCH DATE: April 25, 2006
ORBIT: 500 km (311 mi), 97.4° inclination (Sun-synchronous)
LAUNCH SITE: Svobodny, Russia
LAUNCHER: Start-1
LAUNCH MASS: 350 kg (770 lb)
BODY DIMENSIONS: 2.3 x 4 m (7.5 x 13.1 ft)
PAYLOAD: Panchromatic Imaging Camera-2 (PIC-2)

ERS (European Remote Sensing) EUROPE

Earth remote sensing by radar

ESA's ERS satellites were the first of a new generation of remote sensing satellites, using microwave sensors to obtain all-weather, 24-hour global measurements of wind speed and direction, wave height, surface temperatures, surface altitude, cloud cover, and atmospheric water vapor levels.

ERS-1 was launched on July 17, 1991, followed four years later by ERS-2. The two satellites operated in tandem from August 1995 to May 1996, when the orbits were phased for one-day revisits. ERS-1 was then kept as a back-up until computer and gyro control failure led to its shutdown on March 10, 2000. ERS-1 made 45,000 orbits and acquired over 1.5 million Synthetic Aperture Radar (SAR) scenes.

The satellites have a single 11.7 m (38.4 ft) long solar panel, which provides 2.6k W of power, and a 10 m (33 ft) long SAR antenna. ERS-2 carried the same instruments as ERS-1, with the addition of GOME. It continues to send back data, though it switched to Earth-sensor and reaction-wheel control after the last gyro failed in January 2001.

SPECIFICATION (ERS-2):

MANUFACTURER: Dornier, Matra (now EADS Astrium)
LAUNCH DATE: April 21, 1995
ORBIT: 784 km (487 mi), 98.6° inclination (Sun-synchronous)
LAUNCH SITE: Kourou, French Guiana
LAUNCHER: Ariane 40
LAUNCH MASS: 2516 kg (5535 lb)
BODY DIMENSIONS: 2 x 2 x 3 m (6.6 x 6.6 x 9.8 ft)
PAYLOAD: Active Microwave Instrument (AMI), included a Synthetic Aperture Radar (SAR), wave scatterometer, and wind scatterometer; Radar Altimeter (RA); Along-Track Scanning Radiometer and Microwave Sounder (ASTR-M); Global Ozone Monitoring Experiment (GOME); Laser Retroflector (LRR); Precision Range/Range Rate Experiment (PRARE)

FAST (Fast Auroral Snapshot) USA

Aurora studies

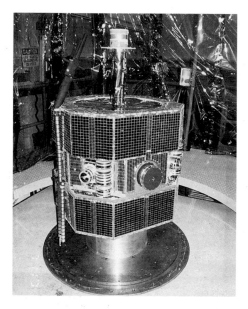

SPECIFICATION:

MANUFACTURER: NASA-GSFC
LAUNCH DATE: August 21, 1996
ORBIT: 350 x 4175 km (217 x
2594 mi), 83° inclination
LAUNCH MASS: 191 kg (420 lb)
LAUNCH SITE: Vandenberg,
California (air launch)
LAUNCHER: Pegasus XL
BODY DIMENSIONS: 1.8 x 1.2 m
(5.9 x 3.9 ft)
PAYLOAD: Electrostatic Analyzer
(ESA); electric field sensors;
Time of Flight Energy Angle
Mass Spectrometer (TEAMS);
AC/DC magnetometers

FAST is the second mission in NASA's Small Explorer Program. The spacecraft was designed to answer fundamental questions about the causes and nature of the Earth's aurora. The satellite crosses Earth's auroral zones four times during each highly elliptical orbit, measuring electric and magnetic fields, plasma waves, energetic electrons and ions, ion mass composition, and thermal plasma density and temperature.

The spacecraft was air-launched from a Lockheed L-1011 aircraft. It has body-mounted solar arrays and is spin-stabilized, rotating at 12 rpm. There are two magnetometer booms on opposite sides, two axial electric field booms and four equally spaced, radial wire electric field booms. It was designed and built by NASA's Goddard Space Flight Center. Instruments were provided by the University of California, Berkeley, with contributions from the University of California, Los Angeles; the University of New Hampshire; and Lockheed Martin. The electric field instrument failed in 2002. FAST was designed for a one-year mission because of anticipated radiation damage, but it remains operational after more than 10 years and 40,000 orbits. By the end of 2006, it had collected over five terabytes of data.

Fengyun-2 (FY-2) CHINA

Meteorology

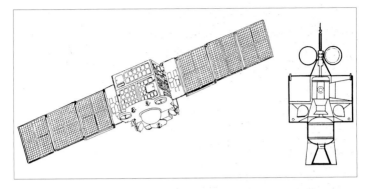

By the beginning of 2007, China had launched eight Fengyun (Wind and Cloud) meteorological satellites: four in the FY-1 series and four in the FY-2 series. The odd-numbered satellites are polar orbiting, while the even-numbered satellites are in geostationary orbit. They are used for weather forecasting and natural-disaster and environmental monitoring.

The FY-1 satellites are box-shaped spacecraft with two solar arrays. Four were launched from Taiyuan by Long March 4 vehicles between 1988 and 2002. The geostationary FY-2 series are cylindrical with body-mounted solar cells. Two experimental spacecraft, FY-2A and FY-2B, were launched in June 1997 and June 2000. The first operational version, FY-2C, was launched in October 2004, followed by FY-2D on December 8, 2006. Together, these two satellites monitor most of Asia, the Indian Ocean, and the West Pacific. Four more FY-2 spacecraft are planned.

The second-generation Fengyun-3 is expected to be launched in the second half of 2007. The three-axis-stabilized FY-3 will have a mass of 2450 kg (5390 lb) and carry new all-weather, multispectral, three-dimensional sensors.

SPECIFICATION (FY-2D):

MANUFACTURER: Shanghai Academy of Space Flight Technology
LAUNCH DATE: Dec. 8, 2006
ORBIT: 86.5° E (GEO)
LAUNCH MASS: 1369 kg (3012 lb)
LAUNCH SITE: Xichang, China
LAUNCHER: Long March 3A
BODY DIMENSIONS: 1 x 1.6 m (3.3 x 5.2 ft)
PAYLOAD: Scanning radiometer; S-band and UHF data transmitter; water vapor sensor

FormoSat-2/ROCSat-2 (Republic of China Satellite-2)

TAIWAN

Earth remote sensing and sprite detection

SPECIFICATION:

MANUFACTURER: EADS Astrium
LAUNCH DATE: May 20, 2004
ORBIT: 891 km (554 mi),
99.1° inclination
LAUNCH SITE: Vandenberg,
California
LAUNCHER: Taurus XL
LAUNCH MASS: 746 kg (1681 lb)
BODY DIMENSIONS: 1.6 x 2.4 m
(5.2 x 7.9 ft)
PAYLOAD: Remote Sensing
Instrument (RSI); Imager of
Sprites: Upper Atmospheric
Lightning (ISUAL)

ROCSat-2 (Republic of China Satellite-2) is the second high-resolution Earth-observation satellite of the Taiwanese National Space Program Office (NSPO). Its main mission is regional remote sensing—collecting data to be used for natural disaster evaluation, agricultural applications, urban planning strategy, environmental monitoring, and ocean surveillance.

Prime contractor EADS Astrium SAS supplied the platform (the first use of the Leostar-500-XO bus) and the RSI remote sensing instrument—the first all silicon carbide instrument. Taiwan provided the satellite computers, S-band antennas and Sun sensors. The structure consists of a hexagonal body. The satellite is three-axis-stabilized. The upper deck of the platform carries the payload and the star sensors and gyroscopes. The lower deck carries four reaction wheels and the propulsion module. The pointing accuracy is less than 0.7 km (2296 ft or 0.12°); the position knowledge is less than 70 m (230 ft or 0.02°). The fixed solar array uses gallium arsenide cells.

ROCSat-2 has a 0.6 m (2 ft) diameter telescope with a 2 m (6.6 ft) resolution black-and-white imager and an 8 m (26.2 ft) resolution color imager, as well as a detector to study lightning "sprites." ROCSat-2 was renamed FormoSat-2 in December 2004.

FormoSat-3/ROCSat-3 (Republic of China Satellite-3)/ COSMIC (Constellation Observing System for Meteorology, Ionosphere, and Climate) TAIWAN

Atmospheric research

SPECIFICATION:

MANUFACTURER: Orbital Sciences
LAUNCH DATE: April 14, 2006
ORBIT: 700 km (435 mi), 72°
inclination (6 planes spaced
24° apart)
LAUNCH MASS (x 6): 70 kg
(154 lb)
LAUNCH SITE: Vandenberg,
California
LAUNCHER: Minotaur I
BODY DIMENSIONS: 1 x 0.2 m
(3.3 x 0.6 ft)
PAYLOAD: Global Positioning
System (GPS) occultation
receiver; ionospheric
photometer; tri-band beacon

The FormoSat-3 program (previously known as ROCSat-3) is an international collaboration between the National Space Organization of Taiwan and the University Corporation for Atmospheric Research in the USA. In the US it is known as COSMIC. It includes a constellation of six microsatellites to collect atmospheric data for weather prediction and for ionosphere, climate, and gravity research.

Orbital Sciences was responsible for constellation design and analysis, development of the spacecraft bus, payload-instrument development, and oversight of components made in Taiwan. Orbital also provided assistance with satellite integration and testing in Taiwan, in-orbit checkout, and mission operations. The identical spacecraft are based on Orbital's MicroStar bus and the OrbView-1 satellite that was launched in 1995. On either side of the thin, cylindrical body are two circular solar arrays.

Each day, their three scientific instruments collect some 2500 atmospheric temperature measurements and water-vapor profiles from ground level to the ionosphere. They also convert GPS radio occultation measurements into a precise worldwide set of weather and climate-change data. Design life is two years.

GOES (Geostationary Operational Environmental Satellites) USA

Meteorology

SPECIFICATION
(GOES-13/GOES N):

MANUFACTURER: Boeing Space
and Intelligence Systems
LAUNCH DATE: May 24, 2006
ORBIT: 105° W (GEO)
LAUNCH MASS: 3133 kg (6893 lb)
LAUNCH SITE: Cape Canaveral,
Florida
LAUNCHER: Delta IV
BODY DIMENSIONS: 4.2 x 1.9 m
(13.8 x 6.2 ft)
PAYLOAD: Solar X-ray imager;
sounder; space environment
monitor; data-collection system;
search and rescue transponder

GOES is a joint weather satellite program involving NASA and the US National Oceanographic and Atmospheric Administration (NOAA). NOAA owns and operates the satellites; NASA is responsible for spacecraft and instrument procurement, design, and development, plus launch.

The GOES space segment comprises two geostationary spacecraft, one located at 75° W and one at 135° W (also known as GOES-East and GOES-West). One monitors North and South America and most of the Atlantic, the other North America and the Pacific Ocean. The first GOES was launched in 1975, followed by a second in 1977; 11 further GOES spacecraft have been launched since then. The satellites are given alphabetic designations prior to launch and are numbered after launch. GOES D–H and N–P were manufactured by Boeing and GOES A–C and I–M by Space Systems/Loral.

The second-generation GOES N–P are based on the Boeing 601 three-axis-stabilized platform. GOES-12 and -13 (GOES M and N) carry a solar X-ray imager in addition to the imager and sounder instruments. In mid-2007, GOES-10 was at 135° W and GOES-12 at 75° W. GOES-13 was an on-orbit back-up. Design life is at least five years.

GRACE (Gravity Recovery and Climate Experiment)

GERMANY/USA

Earth gravity studies

SPECIFICATION:

MANUFACTURER: EADS Astrium/
Space Systems/Loral
LAUNCH DATE: March 17, 2002
ORBIT: 485 km (301 mi),
89° inclination
LAUNCH MASS (x 2): 432 kg
(950 lb)
LAUNCH SITE: Plesetsk, Russia
LAUNCHER: Rockot/Breeze KM
BODY DIMENSIONS: 2 x 3.1 x
0.7 m (6.6 x 10.2 x 2.3 ft)
PAYLOAD: Accelerometer (ACC);
K-Band Ranging System (KBR);
GPS space receiver; Laser
Retroflector (LRR); Star Camera
Assembly (SCA); Coarse Earth
and Sun Sensor (CES); Ultra-
stable Oscillator (USO); Center of
Mass Trim Assembly (CMT)

GRACE is a US–German dual satellite geodetic mission, the first launch in NASA's Earth System Science Pathfinder program. Its overall objective is to measure Earth's gravity field with unprecedented accuracy.

The two identical satellites are trapezoidal in cross section and based on the Astrium FLEXBUS, using heritage hardware from the German CHAMP mission. They differ only in the S-band radio frequencies used for ground communication and in the K-band frequencies used for the intersatellite link. The two main solar arrays are canted to the equipment panel, with two more solar arrays on the "roof." The body is closed at the front and rear by sandwich panels. The aft panel carries the occultation GPS antenna; the front panel has the cutout for the Ku/Ka-band horn. The nadir and zenith S-band antennas are mounted on brackets to minimize gain disturbance.

Flying in formation, their precise speed and separation distance are constantly measured by a microwave ranging instrument. As the gravitational field changes, the distance between them changes. The delay or inclination angle of GPS measurements is used to study atmospheric and ionospheric effects. End of life is expected in 2011.

Himawari (Geostationary Meteorological Satellite, GMS)

JAPAN

Meteorology

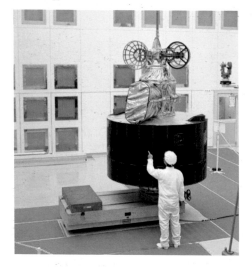

SPECIFICATION
(Himawari-5/GMS-5):

MANUFACTURER: Nippon Electric
Company/Hughes Space and
Communications Company
LAUNCH DATE: March 18, 1995
ORBIT: 140° E (GEO)
LAUNCH SITE: Tanegashima, Japan
LAUNCHER: H-II 3F
LAUNCH MASS: 746 kg (1644 lb)
BODY DIMENSIONS: 2.2 x 3.5 m
(7.1 x 11.6 ft)
PAYLOAD: Visible and Infrared
Spin Scan Radiometer (VISSR);
search and rescue relay

The Japanese Geostationary Meteorological Satellite (GMS) series, also known as Himawari (Sunflower), is operated by the Japan Meteorological Agency. From geostationary orbit at around 140° E, they were used for weather forecasting over the Asian Pacific. GMS was also an important link in the World Weather Watch network.

The first satellite was launched from Cape Canaveral on July 14, 1977. The principal instrument was a US-built radiometer, which was mounted on the spinning spacecraft body and rotated at 100 rpm. The antennas were despun and remained pointed toward Earth. The spin of the satellite enabled a west–east scan, while a north–south scan was achieved by moving the scan mirror.

GMS-2, -3, and -4 also carried a sensor that investigated the solar wind. The last in the series, Himawari-5/GMS-5, carried an experimental payload to relay UHF search-and-rescue data from aircraft and ships. After the 1999 launch failure of its replacement, MTSAT-1, observations of the southern hemisphere were reduced in order to cut fuel consumption. Himawari-5's service life ended on May 22, 2003, when it was temporarily replaced by the US GOES-9. The replacement MTSATs are also known as Himawari.

ICESat (Ice, Cloud, and Land Elevation Satellite) USA

Earth remote sensing

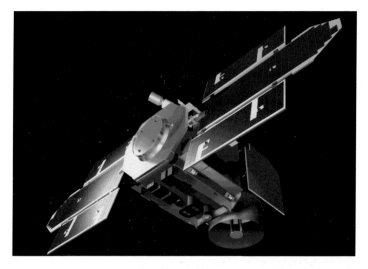

ICESat is part of NASA's Earth Observing System. Its primary goals are to measure changes in ice sheet mass and to understand how changes in the Earth's atmosphere and climate affect polar ice masses and global sea level.

ICESat is based on the Ball Commercial Platform 2000 (BCP 2000) that was also used for the QuikSCAT and QuickBird missions. Two solar arrays provide 640 W of power. It is capable of precision pointing control and can rapidly orient within 5° of the ground track. A solid-state recorder provides 56 Gb of payload data storage, or 24 h of science data. The payload includes the GLAS instrument, two BlackJack GPS receivers, and a star-tracker attitude determination system. After the failure of laser 1 on March 29, 2003, measurements began using laser 2 in autumn 2003. The orbit has a ground-track repeat cycle of 91 days. It is controlled on-orbit by the University of Colorado. Mission life is expected to be at least five years.

SPECIFICATION:

MANUFACTURER: Ball Aerospace
LAUNCH DATE: January 12, 2003
ORBIT: 593 x 610 km (368 x 379 mi), 94° inclination
LAUNCH MASS: 958 kg (2108 lb)
LAUNCH SITE: Vandenberg, California
LAUNCHER: Delta II 7320
BODY DIMENSIONS: 1.9 x 1.9 x 3.1 m (6.2 x 6.2 x 10.2 ft)
PAYLOAD: Geoscience Laser Altimeter System (GLAS); 2 BlackJack GPS receivers

IKONOS USA

Earth remote sensing

SPECIFICATION (IKONOS-2):

MANUFACTURER: Lockheed Martin Space Systems
LAUNCH DATE: September 24, 1999
ORBIT: 681 x 709 km (423 x 441 mi), 98.1° inclination (Sun-synchronous)
LAUNCH MASS: 817 kg (1794 lb)
LAUNCH SITE: Vandenberg, California
LAUNCHER: Athena 2
BODY DIMENSIONS: 1.8 x 1.6 m (6 x 5.1 ft)
PAYLOAD: Optical Sensor Assembly (1 m panchromatic camera and 4 m multispectral camera); GPS receiver

IKONOS was the world's first commercial satellite to collect black-and-white images with 1 m (3.3 ft) resolution, and multispectral imagery with 4 m (13.1 ft) resolution. Data can also be merged to create 1 m (3.3 ft) color images. Both sensors have a swath width of 12 km (7.4 mi). Revisit time is approximately three days, and the local crossing time at the equator is 10.30 am on the descending node. The satellite is owned and operated by GeoEye (formerly Space Imaging Inc.). The name IKONOS comes from the Greek word for "image."

IKONOS-1 was lost in April 1999 when the Athena 2 payload shroud failed to separate. The second satellite was launched five months later. IKONOS-2 is a three-axis-stabilized spacecraft based on the hexagonal Lockheed Martin LM900 bus. 1.5 kW of power is provided by three top-mounted solar panels. Attitude is determined by two star trackers and a sun sensor and altered by means of four reaction wheels. Location is determined with the aid of a GPS receiver. The data-recording capacity of the onboard solid-state memory is 64 Gb. Data are transmitted in X-band to dedicated ground stations around the world. The satellite is expected to operate until 2008 or beyond.

IMAGE (Imager for Magnetopause-to-Aurora Global Exploration) USA

Magnetosphere/aurora imaging

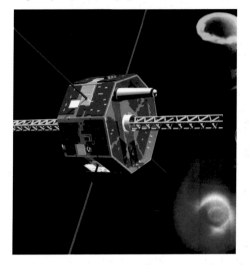

SPECIFICATION:

MANUFACTURER: Lockheed Martin
LAUNCH DATE: March 25, 2000
ORBIT: 45,922 x 1000 km
(28,535 x 620 mi), 90.01°
inclination
LAUNCH MASS: 494 kg (1087 lb)
LAUNCH SITE: Vandenberg,
California
LAUNCHER: Delta II 7326
BODY DIMENSIONS: 2.3 x 1.5 m
(7.4 x 4.5 ft)
PAYLOAD: Low-, medium- and
high-energy neutral atom
imagers (LENA, MENA, HENA);
Far-Ultraviolet (FUV) Imaging
System; Extreme Ultraviolet
(EUV) Imager; Radio Plasma
Imager (RPI); Central Instrument
Data Processor (CIDP)

IMAGE was NASA's first Medium-class Explorer, and the first spacecraft dedicated to imaging Earth's magnetosphere. It used neutral-atom, ultraviolet, and radio-imaging techniques to produce the first global images of plasmas in the inner magnetosphere. The real-time data were used by Japan and the US National Oceanic and Atmospheric Administration for space weather forecasts.

IMAGE was an octagonal spin-stabilized spacecraft. Dual-junction gallium arsenide solar arrays attached to the spacecraft's side and end panels provided power. The instruments were located on a payload deck in the middle of the spacecraft. Subsystems for electrical power, communications, command and data handling, and attitude determination and control were mounted in four bays below the payload deck. Cutouts in the side panels accommodated the instrument apertures, the deployers for the RPI radial antennas, and radiators used for thermal control. The RPI used two 10 m (32.8 ft) axial antennas above and below the spacecraft, plus four 250 m (820 ft) wire radial antennas located 90° apart. Spin period was 2 min (0.5 rpm).

Three S-band antennas provided communication with the ground—a medium-gain helix antenna and two low-gain omnidirectional antennas. IMAGE continued operating until December 18, 2005, when its power supply subsystems failed.

IRS (Indian Remote Sensing) INDIA

Earth remote sensing

SPECIFICATION (IRS-P3):

MANUFACTURER: ISRO
LAUNCH DATE: March 21, 1996
ORBIT: 817 km (508 mi), 98.7°
inclination (Sun-synchronous)
LAUNCH SITE: Sriharikota, India
LAUNCHER: PSLV-D3
LAUNCH MASS: 922 kg (2028 lb)
BODY DIMENSIONS: 1.6 x 1.6 x
1.1 m (5.2 x 5.2 x 3.6 ft)
PAYLOAD: Wide Field Sensor
(WiFS); Multispectral
Optoelectronic Scanner (MOS);
X-Ray Astronomy Payload (XRAP)

The IRS series was India's first domestic program of dedicated Earth-resources satellites. The first satellite, IRS-1A, was launched by a Vostok on March 17, 1988; the next two missions of the IRS-1 series were also launched on Russian rockets. The fourth and final satellite of the series, IRS-1D, was launched by an Indian PSLV and erroneously placed in a highly elliptical orbit. These were followed by the pre-operational IRS-P series, which were intended for technology evaluation and all launched by the PSLV.

The box-shaped IRS-P3 spacecraft structure consists of four vertical panels and two horizontal decks. The payload is on the upper deck. Two Sun-tracking solar panels provide 873 W of power. The three-axis-stabilized spacecraft uses Earth sensors, Sun sensors, and gyros as attitude sensors; it also has reaction wheels, magnetic torquers, and a system of thrusters. The MOS instrument was provided by DLR (German Aerospace Center).

The X-ray observations ended on June 30, 2000, because of depletion of the propellant needed for pointed observations. IRS-P3 encountered power problems in 2003, and its orbit began to alter. Data transmissions from MOS came to an end on May 31, 2004, and all operations ended a few months later. The most recent IRS satellites have been given names (e.g. CARTOSAT), and are discussed in separate entries.

Jason-1 USA/FRANCE
Ocean altimetry observation

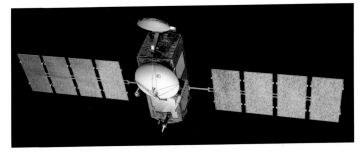

Jason-1 is a joint program involving NASA and the French space agency CNES. Its mission was to improve our understanding of ocean circulation and its effect on global climate. The primary objectives included measuring ocean-surface topography and global sea-level change.

As the follow-on to Topex/Poseidon, the spacecraft retains many similar features, although it is five times lighter and was the first to use the French Proteus bus. The three-axis-stabilized satellite is approximately rectangular, with two solar arrays generating 450 W.

Jason-1 was launched together with NASA's TIMED spacecraft. It was placed in an orbit that lagged 1–10 min behind Topex/Poseidon but was otherwise nearly identical, with a repeat cycle of almost 10 days.

Jason-1 carries five instruments. Poseidon-2 is the main instrument, used to measure range. The JMR measures perturbations due to atmospheric water vapor. Three location systems, Doris, LRA, and TRSR, allow precise determination of the orbit. NASA has responsibility for satellite control and its instruments. CNES operates the data-processing center. The mission was designed to last three years. Jason-2 is due for launch in 2008.

SPECIFICATION:

MANUFACTURER: Alcatel Space
LAUNCH DATE: December 7, 2001
ORBIT: 1336 km (830 mi), 66° Inclination
LAUNCH MASS: 500 kg (1102 lb)
LAUNCH SITE: Vandenberg, California
LAUNCHER: Delta II 7920
BODY DIMENSIONS: 1.9 x 1.9 x 3.3 m (6.2 x 6.2 x 10.8 ft)
PAYLOAD: Poseidon-2 radar altimeter; Jason Microwave Radiometer (JMR); Doppler Orbitography and Radio-positioning Integrated by Satellite (DORIS); Turbo Rogue Space Receiver (TRSR); Laser Retroflector Array (LRA)

Kalpana (MetSat) INDIA

Meteorology

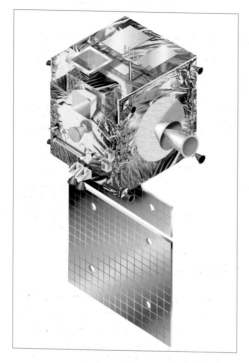

SPECIFICATION:

MANUFACTURER: ISRO Satellite Center
LAUNCH DATE: September 12, 2002
ORBIT: 74° E (GEO)
LAUNCH SITE: Sriharikota, India
LAUNCHER: PSLV-C4
LAUNCH MASS: 1055 kg (2321 lb)
PAYLOAD: Very-High-Resolution Radiometer (VHRR); Data Relay Transponder (DRT)

MetSat-1 is the first dedicated GEO weather-satellite project built by ISRO; previously, meteorological services had been combined with telecommunication and television services in the INSAT series. Launched into GTO by the PSLV, MetSat-1's orbit was later raised to the final GEO position by firing the satellite's liquid apogee motor. On February 5, 2003, MetSat-1 was renamed Kalpana-1 to honor the Indian-born astronaut Kalpana Chawla, who died in the *Columbia* STS-107 accident.

Kalpana is three-axis stabilized and was designed using the new I-1000 spacecraft bus, which includes lightweight structural elements such as carbon-fiber reinforced plastic. Eight thrusters are used for orbit and attitude control. The box-shaped satellite has a lightweight planar-array antenna that transmits data to the ground. It has two high-capacity magnetic torquers to overcome the solar-radiation pressure effects. The single gallium arsenide solar array measures 2.2 x 1.9 m (6.6 x 6.1 ft) and generates 550 W of power.

Kompass 2/COMPASS 2 (Complex Orbital Magneto-Plasma Autonomous Small Satellite 2) RUSSIA

Ionosphere studies

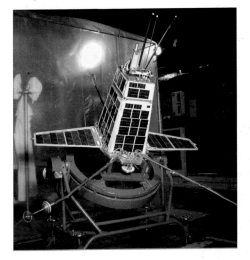

SPECIFICATION:

MANUFACTURER: IZMIRAN, Makeyev Federal Rocket Center
LAUNCH DATE: May 26, 2006
ORBIT: 399 x 483 km (248 x 300 mi), 78.9° inclination (Sun-synchronous)
LAUNCH MASS: 77 kg (170 lb)
LAUNCH SITE: Barents Sea (*Ekaterinburg* nuclear missile submarine)
LAUNCHER: Shtil 1
BODY DIMENSIONS: 1.7 x 0.8 x 0.5 m (3.7 x 2.6 x 1.6 ft)
PAYLOAD: Radiation and ultraviolet detectors (Tatyana); Radio Frequency Analyzer (RFA); Low-frequency wave complex (NVK); Dual-frequency radio transmitter (MAYAK)

The Kompass minisatellites are pilots for a multisatellite system that will be used for disaster prediction and monitoring. The spacecraft are developed as part of the Russian federal space program. Contractors include the R&D Electromechanical Institute, the Institute of Terrestrial Magnetism, Ionosphere and Radiowave Propagation (IZMIRAN), and the Makeyev state rocket center.

The Kompass spacecraft is three-axis stabilized. The bus has the shape of an inverted pyramid. Two deployable solar panels produce 50 W of power. The overall objective of Kompass 2 is to demonstrate sounding techniques that may eventually lead to the forecasting of natural disasters such as earthquakes. The payload includes Tatyana, developed by the Skobeltsyn Institute of Nuclear Physics of Moscow State University, which is mainly intended to monitor the ionosphere and detect electromagnetic signatures associated with earthquakes and volcanic eruptions.

Kompass 1 was launched as a Zenit-2 secondary payload on December 10, 2001, but it went out of control and was lost shortly after deployment. After launch from a nuclear missile submarine in May 2006, contact was lost with Kompass 2 until November 16, and first data were received on November 25, 2006. It was declared operational in early March 2007. Design life is five years.

Kompsat (Korea Multipurpose Satellite)/Arirang

SOUTH KOREA

Earth remote sensing

SPECIFICATION
(Kompsat-2):

MANUFACTURER: Korea Aerospace
Research Institute
LAUNCH DATE: July 28, 2006
ORBIT: 685 km (426 mi),
98.1° inclination
(Sun-synchronous)
LAUNCH MASS: 798 kg (1756 lb)
LAUNCH SITE: Plesetsk, Russia
LAUNCHER: Rockot
BODY DIMENSIONS: 1.9 x 2.6 m
(6.1 x 8.5 ft)
PAYLOAD: Multi-spectral camera

The civil–military dual-purpose Kompsat program began with the launch of Kompsat-1 in December 1999. Built in collaboration with the USA, the spacecraft was based on TRW's Eagle class of lightweight, modular spacecraft. Its payload included a 6.6 m (21.6 ft) ground resolution CCD imaging system supplied by Litton Itek Optical Systems, while TRW supplied a low-resolution camera for ocean and Earth-resources monitoring. Also onboard was an instrument to measure Earth's ionospheric and magnetic fields and a high-energy particle detector.

Kompsat-2 is the first high-resolution optical reconnaissance satellite developed by South Korea. It uses a similar design to Kompsat-1, with an optical imager provided by Israel. This is capable of detecting objects 1 m (3.3 ft) in diameter in black-and-white mode and has 4 m (13.1 ft) resolution in multispectral mode.

Kompsat-3, which will be similar to Kompsat-2, is tentatively scheduled for launch in 2009. South Korea's long-term space program includes two more Earth-observation satellites. Kompsat-5, which will carry a synthetic-aperture radar imager provided by Alcatel Alenia Space, is scheduled for launch in 2008. This radar imager will have a ground resolution of 1–3 m (3.3–10 ft).

Landsat USA

Earth remote sensing

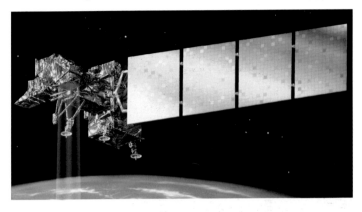

The first Landsat—originally called ERTS (Earth Resources Technology Satellite)—was developed and launched by NASA in 1972. The spacecraft was derived from the Tiros-N and DMSP weather satellites. Six more Landsats have been launched since then, providing 35 years of imaging of Earth's land masses.

Landsats 4 and 5 carried a Thematic Mapper (TM) and a Multispectral Scanner (MSS) radiometer. An Enhanced Thematic Mapper (ETM) was introduced with Landsat 6 in 1993, but the satellite was lost due to a ruptured hydrazine manifold. Landsat 7 also carries ETM, which gives 30 m (100 ft) spatial resolution in the seven spectral bands and 15 m (49 ft) resolution in black and white. The Landsat 7 ETM+ experienced a failure of its Scan Line Corrector in May 2003, causing gaps in data coverage. The ground-track repeat time is 16 days. Improvements over previous versions include more precise instrument calibration and a solid-state recorder capable of storing 100 high-resolution images. Power is provided by a single Sun-tracking solar array. The US Geological Survey took over operation of the Landsats from NASA in 2000.

SPECIFICATION
(Landsat 7):

MANUFACTURER: Lockheed Martin Space Systems
LAUNCH DATE: April 15, 1999
ORBIT: 702 km (436 mi), 98.2° inclination (Sun-synchronous)
LAUNCH MASS: 1969 kg (4340 lb)
LAUNCH SITE: Vandenberg, California
LAUNCHER: Delta II 7920
BODY DIMENSIONS: 4 x 2.7 m (13.3 x 9 ft)
PAYLOAD: Enhanced Thematic Mapper (ETM+)

Meteor RUSSIA

Meteorology

SPECIFICATION
(Meteor-3M-1):

MANUFACTURER: VNIIEM (All-Russian Scientific and Research Institute of Electromechanics)
LAUNCH DATE: December 10, 2001
ORBIT: 996 x 1015 km (618 x 630 mi), 99.7° inclination (Sun-synchronous)
LAUNCH SITE: Baikonur, Kazakhstan
LAUNCHER: Zenit-2
LAUNCH MASS: 2500 kg (5500 lb)
BODY DIMENSIONS: 1.4 x 2.2 m (6.4 x 7.2 ft)
PAYLOAD: MR-2000 TV scanner; infrared scanner (ART); microwave radiometer (MIVZA); microwave radiometer (MTVZA); particle instrument (KGI-4); Measurement of Geoactive Emissions (MSGI-5); ultraviolet spectrometer (SFM-2); multichannel scanning device (MSU-E); multichannel scanning device (MSU-S); Stratospheric Aerosols and Gas Experiment (SAGE III)

Meteor is the Russian generic name for the family of meteorological satellites that fly in LEO. They were developed on behalf of ROSHYDROMET, the Russian Federal Service for Hydrometeorology and Environmental Monitoring.

The first experimental weather satellites were given the Cosmos label, starting with Cosmos-44, which was launched on August 25, 1964. Nine more Cosmos weather satellites were launched until 1969, when Meteor-1 was launched. Meteor-2, first flown in 1975, introduced the new operational orbit of 950 km (590 mi), 82.5° inclination. Meteor-3 followed in 1985. In 2001, the Meteor-3M-1 satellite was the first to be placed in a Sun-synchronous orbit.

The Meteor-3M spacecraft was an advanced version, with a mass almost twice that of Meteor-2, improved stabilization, and a larger payload of up to 900 kg (1980 lb). The bus was cylindrical with two solar panels providing about 1 kW of power. The payload included NASA's SAGE III ozone monitor and other instruments designed to measure temperature and humidity profiles, clouds, surface properties, and high-energy particles in the upper atmosphere. It was in a Sun-synchronous orbit with an ascending node time of about 9.00 am. Design lifetime was three years. The Meteor-3M-1 spacecraft stopped functioning on April 5, 2006 due to a power supply system failure.

Meteosat/MSG EUROPE

Meteorology

SPECIFICATION (MSG-2):

MANUFACTURER: Alcatel Space
LAUNCH DATE: December 21, 2005
ORBIT: 0° (GEO)
LAUNCH MASS: 2034 kg (4484 lb)
LAUNCH SITE: Kourou, French Guiana
LAUNCHER: Ariane 5G
BODY DIMENSIONS: 3.2 x 3.7 m (10.5 x 12.1 ft)
PAYLOAD: Spinning Enhanced Visible and Infrared Imager radiometer (SEVIRI); Geostationary Earth Radiation Budget (GERB)

Meteosat is Europe's series of geostationary weather satellites. Their primary goal is to provide visible and infrared day/night images of cloud cover and temperature. They provide data every 30 min from three spectral channels (visible, infrared, and water vapor). Designed by ESA and operated by the European Organization for the Exploitation of Meteorological Satellites (EUMETSAT), they have been operational since 1977. The Meteosats have provided images of the Europe–Africa hemisphere and data for weather forecasts for a quarter of a century. The last of the series (Meteosat-7) followed in 1997 and is still operational.

The Meteosat Second Generation (MSG) system is a significantly improved follow-on that will comprise four geostationary satellites that will operate consecutively until 2018. The drum-shaped satellite is coated with solar cells producing 750 W of power. The spin-stabilized spacecraft rotates at 100 rpm. Its key instrument is the SEVIRI radiometer, which provides images with a resolution of 3 km (1.9 mi) in the infrared and 1 km (0.6 mi) at visible wavelengths. The GERB instrument measures the Earth's radiation balance. Data transmission is almost 20 times faster (up to 3.2 Mbps) than earlier Meteosats. Design life is seven years.

MetOp EUROPE

Meteorology

SPECIFICATION (MetOp-A):

MANUFACTURER: EADS-Astrium
LAUNCH DATE: October 19, 2006
ORBIT: 817 km (508 mi), 98.7°
inclination (Sun-synchronous)
LAUNCH MASS: 4086 kg (8989 lb)
LAUNCH SITE: Baikonur,
Kazakhstan
LAUNCHER: Soyuz 2-1A
BODY DIMENSIONS: 6.2 x 3.4 x
3.4 m (20.3 x 11.2 x 11.2 ft)
PAYLOAD: Infrared Atmospheric
Sounding Interferometer (IASI);
Microwave Humidity Sounder
(MHS); Global Navigation
Satellite System Receiver for
Atmospheric Sounding (GRAS);
Advanced Scatterometer (ASCAT);
Global Ozone Monitoring
Experiment-2 (GOME-2);
Advanced Microwave Sounding
Units (AMSU-A1 and AMSU-A2);
High-resolution Infrared
Radiation Sounder (HIRS/4);
Advanced Very-High-Resolution
Radiometer (AVHRR/3);
Advanced Data-Collection
System (A-DCS or Argos); Space
Environment Monitor (SEM-2);
Search and Rescue Processor
(SARP-3); Search and Rescue
Repeater (SARR)

MetOp is the basis of the EUMETSAT Polar System (EPS), Europe's first polar-orbiting operational meteorological satellite system. It is the European contribution to the Initial Joint Polar-Orbiting Operational Satellite System with the US National Oceanic and Atmospheric Administration (NOAA). EPS consists of three MetOp satellites, together with their ground system, with an operational life of at least 14 years.

The first MetOp replaced one of two polar orbiting satellites operated by NOAA. Its payload includes five new-generation instruments developed by Eumetsat, ESA and CNES, in addition to instruments already flown on NOAA satellites. These provide a wide range of atmospheric data as well as images of clouds, land and ocean surfaces. MetOp flies in a polar orbit corresponding to local "morning" while the US is responsible for "afternoon" coverage. The satellite is based on ESA's Polar Platform, also used for Envisat. It has a single, detached solar array, providing 3.9 kW. The box-shaped service module includes equipment for command and control, communications with the ground, power and orbit control, and propulsion. Design life is five years.

NOAA (National Oceanic and Atmospheric Administration) POES (Polar Orbiting Environmental Satellites) USA

Meteorology

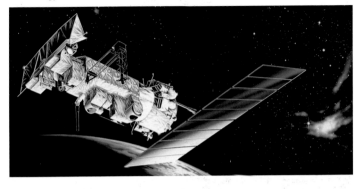

The POES series of polar orbiting weather satellites is a joint program of NASA, NOAA, France, Canada, the UK, and EUMETSAT. Satellites are given a letter prior to reaching orbit, then a number: for example, NOAA-M was redesignated NOAA-17 after reaching LEO.

One satellite typically crosses the equator at 7.30 am local time, the other at 1.40 pm. This ensures that data for any region are no more than 6 h old. With the launch of NOAA-17, the crossing time of the morning series spacecraft was changed to 10.00 am. It was the last NOAA satellite in the morning orbit, which was to be taken over by Europe's MetOp series. NOAA-18 (N), launched May 20, 2005, is the primary satellite for the evening orbit.

The POES fifth generation began with NOAA-K (NOAA-15). They use a Star 37XFP motor to circularize the orbit after separation. A suite of instruments measures many features of the Earth's atmosphere. The primary instrument is the AVHRR, which operates in six visible and infrared bands. Spatial resolution is about 1 km (0.6 mi).

SPECIFICATION (NOAA-17, NOAA-M):

MANUFACTURER: NASA-GSFC
LAUNCH DATE: June 24, 2002
ORBIT: 807 x 822 km (501 x 510 mi), 98.8° inclination (Sun-synchronous)
LAUNCH SITE: Vandenberg, California
LAUNCHER: Titan II
LAUNCH MASS: 2232 kg (4910 lb)
BODY DIMENSIONS: 4.2 x 1.9 x 7.4 m (13.8 x 6.2 x 24.3 ft)
PAYLOAD: Advanced Very-High-Resolution Radiometer (AVHRR-3); High-Resolution Infrared Sounder (HIRS-3); Advanced Microwave Sounding Unit (AMSU-B); Solar Backscatter Ultraviolet Radiometer (SBUV-2); Space Environment Monitor-2 (SEM-2); Argos Data Collection System (DCS-2)

OrbView ^{USA}USA

High-resolution earth remote sensing

SPECIFICATION
(OrbView-3):

MANUFACTURER: Orbital Sciences
LAUNCH DATE: June 26, 2003
ORBIT: 470 km (292 mi), 97.3°
inclination (Sun-synchronous)
LAUNCH MASS: 360 kg (792 lb)
LAUNCH SITE: Vandenberg,
California (air launch)
LAUNCHER: Pegasus XL
BODY DIMENSIONS: 1.9 x 1.2 m
(6.2 x 3.9 ft)
PAYLOAD: 1 m (3.3 ft)
panchromatic camera; 4 m
(13.1 ft) multispectral camera

The OrbView series of Earth-imaging satellites was developed for US company Orbimage (now known as GeoEye). The small, disk-shaped OrbView-1 (Micro-Lab-1) was launched in 1995 and operated until April 2000. OrbView-2 (SeaStar), launched in 1997, delivered ocean color data from NASA's SeaWIFS instrument. OrbView-4 was lost after a Taurus launch failure on September 21, 2001.

OrbView-3 was one of the first commercial satellites to provide high-resolution Earth imagery. Based on the Orbital Sciences LeoStar bus, the three-axis-stabilized, cylindrical spacecraft has a single solar array on top that provides 625 W of power. The bus structure is divided into three modules (propulsion, core, and payload). OrbView-3 has a camera that takes 1 m (3.3 ft) resolution panchromatic (black-and white) and 4 m (13.1 ft) resolution multispectral images with 8 km (5 mi) swath width. Equatorial crossing is at 10.30 am on the descending node. Revisit time is less than three days. On March 4, 2007, the imaging system malfunctioned, and OrbView-3 was declared a total loss on April 23. OrbView-5, now renamed GeoEye-1, will launch in 2008.

PROBA (Project for OnBoard Autonomy) EUROPE

Earth remote sensing

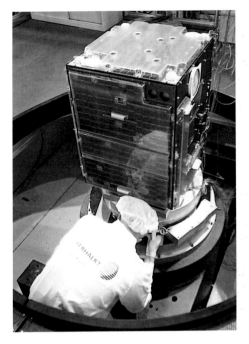

SPECIFICATION:

MANUFACTURER: Verhaert
LAUNCH DATE: October 22, 2001
ORBIT: 550 x 670 km (342 x
416 mi), 98.75° inclination
(Sun-synchronous)
LAUNCH MASS: 149 kg (328 lb)
LAUNCH SITE: Sriharikota, India
LAUNCHER: PSLV-C3
BODY DIMENSIONS: 0.6 x 0.6 x
0.8 m (2 x 2 x 2.6 ft)
PAYLOAD: High-Resolution
Camera (HRC); Compact High-
Resolution Imaging
Spectrometer (CHRIS); Debris
In-orbit Evaluator (DEBIE);
Standard Radiation Environment
Monitoring (SREM);
Miniaturized Radiation Monitor
(MRM); Wide Angle Camera
(WAC)

PROBA is a technology-demonstration mission developed under ESA's General Studies Program to test a platform suitable for small scientific or application missions. The main structure, orbit and attitude control systems, onboard computers, and telemetry are developed and funded by ESA. Four Earth-observation instruments are included in the payload to test platform-pointing and data-management capabilities. The payloads are not funded by ESA but have been offered a free ride on PROBA.

The cube-shaped microsatellite is three-axis stabilized. A novel feature of PROBA-1 is that it can be maneuvered in orbit using a set of four reaction wheels. These allow the satellite to be pointed off-nadir in both the along-track and across-track directions. Gallium arsenide solar arrays mounted on five of the six faces of the spacecraft provide 90 W of power.

The payload is controlled by a computer system 50 times more powerful than its counterpart on SOHO. This enables PROBA to combine onboard mission planning, navigation, and failure detection with some Earth remote sensing. Design life was two years.

Radarsat CANADA

Earth remote sensing by radar

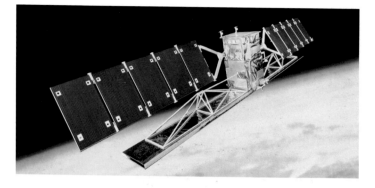

Radarsat is a collaboration between the Canadian government and Canadian/US industry. MacDonald, Dettwiler and Associates own and operate the satellite and ground segment. The Canadian Space Agency helps fund construction and launch in return for images. One of the first satellites to offer commercial radar imagery of the Earth's surface, it enabled Canada to take a lead role in Earth observation.

Radarsat-1 carries a SAR instrument that provides images in all conditions. The satellite has a 15 m (49.2 ft) long SAR antenna and two solar arrays generating 2.5 kW. The SAR transmits a C-band microwave pulse and measures the amount of energy that is reflected back to the satellite. It was the first radar satellite to provide beam steering, with image resolutions varying from 8–100 m (26–328 ft). Design life was five years, but it continued to operate after 12 years.

Radarsat-2 is to be launched from Baikonur in 2007. It will provide C-band SAR data similar to those from Radarsat-1, but it will also include a 3 m (10 ft) high-resolution mode, a full range of signal-polarisation modes, superior data storage, and more precise measurements of satellite position and attitude.

SPECIFICATION (Radarsat-1):

MANUFACTURER: Spar Aerospace (now MacDonald, Dettwiler)/ Ball Aerospace
LAUNCH DATE: November 4, 1995
ORBIT: 793 x 821 km (493 x 510 mi), 98.6° inclination (Sun-synchronous)
LAUNCH SITE: Vandenberg, California
LAUNCHER: Delta II 7920
LAUNCH MASS: 2750 kg (6050 lb)
BODY DIMENSIONS: 1.5 x 1.2 m (4.9 x 3.9 ft)
PAYLOAD: Synthetic Aperture Radar (SAR)

SMOS (Soil Moisture and Ocean Salinity) EUROPE

Studies of soil moisture and ocean salinity

SPECIFICATION:

MANUFACTURER: Alcatel Alenia (now Thales Alenia Space)
LAUNCH DATE: 2008?
ORBIT: 763 km (474 mi), 98.4° (Sun-synchronous)
LAUNCH SITE: Plesetsk, Russia
LAUNCHER: Rockot
LAUNCH MASS: 683 kg (1503 lb)
BODY DIMENSIONS: 2.4 x 2.3 m (7.9 x 7.5 ft)
PAYLOAD: Microwave Imaging Radiometer using Aperture Synthesis (MIRAS)

SMOS is the second Earth Explorer Opportunity mission to be flown under ESA's Living Planet Program. It is designed to provide global maps of soil moisture over the Earth's land masses and of the salinity of the oceans. Soil-moisture data will feed into hydrological studies, while data on ocean salinity will improve our understanding of ocean circulation patterns. Overall, SMOS data will further our knowledge of the Earth's water cycle and improve climate, weather, and extreme-event forecasting.

SMOS is based on the small, cube-shaped Proteus bus developed by CNES and Alcatel Alenia. This acts as a service module accommodating all the major subsystems. The single scientific instrument and its Y-shaped antenna are mounted on the top of the spacecraft through four interface pods. A GPS receiver is used for orbit determination and control, and four hydrazine thrusters are mounted on the base. Two solar arrays provide 900 W of power.

MIRAS is a new type of radiometer that operates between 1400 and 1427 MHz (L-band). The required high resolution is made possible by synthesizing the antenna aperture through multiple small antennae. Sixty-nine antenna elements are spread across the three arms. Each element measures radiation emitted from the Earth. SMOS will fly in a Sun-synchronous dawn–dusk orbit. Design life is three years.

SPOT (Satellite Pour l'Observation de la Terre) FRANCE

Earth remote sensing

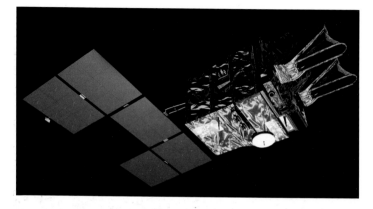

The SPOT Earth-observation satellite system, first approved in 1978, was designed by CNES and developed in cooperation with Sweden and Belgium. Five satellites have been launched since SPOT 1 in 1986. SPOT 2 was placed in orbit in January 1990, followed by SPOT 3 in September 1993 and SPOT 4 in March 1998. The latest is SPOT 5, launched in May 2002. In September 2007, SPOT 2, 4, and 5 were operational.

SPOT 5 uses the same Matra Marconi Space (now Astrium) SPOT Mark 2 extended platform as SPOT 4 and Helios-1. This houses the main subsystems including the DORIS antenna. Protruding at the front end are the two main imaging instruments. Beneath HRG-1 is the Vegetation instrument, first carried on SPOT 4. Alcatel Space (now Thales Alenia) was prime contractor for the Vegetation instrument. The three-axis-controlled spacecraft has a single solar array providing 2.4 kW of power. An onboard solid-state recorder can store 90 Gb of imagery.

SPOT 5 carries two new HRG instruments, derived from the lower-resolution HRVIR instruments on SPOT 4. Design life is five years.

SPECIFICATION:

MANUFACTURER: EADS Astrium
LAUNCH DATE: May 4, 2002
ORBIT: 822 km (511 mi), 98.7° inclination (Sun-synchronous)
LAUNCH SITE: Kourou, French Guiana
LAUNCHER: Ariane 42P
LAUNCH MASS: 3030 kg (6666 lb)
BODY DIMENSIONS: 3.1 x 3.1 x 5.7m (10.2 x 10.2 x 18.7 ft)
PAYLOAD: 2 High Resolution Geometric instruments (HRG); High Resolution Stereoscopic instrument (HRS); Vegetation 2; Doppler Orbitography and Radiopositioning Integrated by Satellite (DORIS)

Terra USA

Earth remote sensing and atmospheric observation

Terra (formerly known as EOS/AM-1) is the flagship of NASA's Earth Observing Satellite program. The project includes collaboration with Japan and Canada. NASA provided the spacecraft, the launch, and three instruments (CERES, MISR, and MODIS); Japan provided ASTER and Canada MOPITT.

Terra is a three-axis-stabilized design with a single rotating, flexible blanket solar array. Data transmissions are via the TDRS system at Ku- and X-band. A high-gain antenna is mounted on the top (zenith) face. A solid-state data recorder can store up to 140 GB.

Terra carries five instruments that simultaneously study clouds, water vapor, aerosol particles, trace gases, and terrestrial and oceanic properties. Every one to two days, the instruments collect data at visible and infrared wavelengths over the Earth's entire surface. Equatorial crossing on the descending node is 10.30 am local time. The design life is six years.

SPECIFICATION:

MANUFACTURER: Lockheed Martin Missiles and Space
LAUNCH DATE: December 18, 1999
ORBIT: 702 km (436 mi), 98.2° inclination (Sun-synchronous)
LAUNCH SITE: Vandenberg, California
LAUNCHER: Atlas-Centaur IIAS
LAUNCH MASS: 5190 kg (11,418 lb)
BODY DIMENSIONS: 6.8 x 3.5 m (22.3 x 11.5 ft)
PAYLOAD: Advanced Spaceborne Thermal Emission and Reflection Radiometer (ASTER); Clouds and the Earth's Radiant Energy System (CERES); Moderate-Resolution Imaging Spectroradiometer (MODIS); Measurements of Pollution in the Troposphere (MOPITT); Multiangle Imaging Spectroradiometer (MISR)

TerraSAR-X GERMANY

Radar remote sensing

TerraSAR-X is an advanced synthetic aperture radar satellite system for scientific and commercial applications. It has been developed under a public/private partnership between the German Aerospace Research Center (DLR) and Astrium.

The TerraSAR-X is based on a hexagonal AstroSat-1000 bus. Its body-mounted solar array provides 800 W of power. The X-band (9.65 GHz) SAR antenna, also side-mounted, gives radar data in various modes. The other nadir-looking side carries an S-band communications antenna, a 3.3 m (10.8 ft) boom for the SAR data-downlink antenna, and a laser retroflector to support precise orbit determination.

The SAR Spot-Light mode yields the highest-resolution data, with 1 m (3.3 ft) pixels for a 10 x 10 km (6.2 x 6.2 mi) image. The ScanSAR mode delivers 16 m (52.5 ft) resolution in a 100 km (62 mi) wide swath. A special split antenna mode allows experimental in-track interferometry—for example, the mapping of moving objects. The satellite flies in a dawn–dusk orbit with an 11-day repeat period. Equatorial crossing on the ascending node is around 6.00 pm. Design life is at least five years.

SPECIFICATION:

MANUFACTURER: EADS Astrium
LAUNCH DATE: June 15, 2007
ORBIT: 514 km (319 mi), 97.4° inclination (Sun-synchronous)
LAUNCH MASS: 1230 kg (2706 lb)
LAUNCH SITE: Baikonur, Kazakhstan
LAUNCHER: Dnepr-1
BODY DIMENSIONS: 5 x 2.4 m (16.4 x 7.9 ft)
PAYLOAD: X-band Synthetic Aperture Radar (TSX-SAR); Laser Communication Terminal (LCT); Tracking, Occultation, and Ranging Experiment (TOR)

TIMED (Thermosphere, Ionosphere, Mesosphere Energetics, and Dynamics) USA

Upper-atmosphere studies

TIMED is the first of NASA's Solar Terrestrial Probes and the first mission ever to conduct a comprehensive, global study of the Earth's mesosphere and lower thermosphere/ionosphere, a poorly understood region of Earth's upper atmosphere. It was launched during solar maximum but has survived to collect data through solar minimum.

TIMED is designed, built, and operated by the Johns Hopkins University Applied Physics Laboratory (APL). The three-axis-stabilized spacecraft is controlled with reaction wheels and torque rods, and uses a ring-laser gyro and two star trackers for attitude estimation. Two solar arrays provide 420 W of power.

Since launch, TIMED has shown how different parts of Earth's atmosphere react to variations in solar activity, including a total solar eclipse in spring 2006. Designed to operate for two years, it was still operating after five. Operations are expected to continue until 2010.

SPECIFICATION:

MANUFACTURER: JHU/APL
LAUNCH DATE: December 7, 2001
ORBIT: 625 km (388 mi), 74.1° inclination
LAUNCH SITE: Vandenberg, California
LAUNCHER: Delta II 7920
LAUNCH MASS: 587 kg (1294 lb)
BODY DIMENSIONS: 2.7 x 1.6 m (8.9 x 5.3 ft)
PAYLOAD: Global Ultraviolet Imager (GUVI); Solar Extreme Ultraviolet Experiment (SEE); TIMED Doppler Interferometer (TIDI); Sounding of the Atmosphere using Broadband Emission Radiometry (SABER)

Topex/Poseidon USA/FRANCE

Ocean altimetry

SPECIFICATION:

MANUFACTURER: Fairchild Space Co.
LAUNCH DATE: August 10, 1992
ORBIT: 1336 km (830 mi), 66° inclination
LAUNCH MASS: 2402 kg (5296 lb)
LAUNCH SITE: Kourou, French Guiana
LAUNCHER: Ariane 42P
BODY DIMENSIONS: 2.8 m x 5.5 x 6.6 m (9.2 x 18 x 21.7 ft)
PAYLOAD: NASA dual-frequency (C- and Ku-band) altimeter; CNES single-frequency (Ku-band) solid-state altimeter; Microwave radiometer; Global Positioning System (GPS) demonstration receiver; laser retroflector; Doppler Orbitography and Radiopositioning Integrated by Satellite (DORIS)

The US–French Topex/Poseidon mission was the follow-on to NASA's Seasat mission. "Topex" is short for "Ocean Topography Experiment," the name of the original US mission proposal, while Poseidon was the name of the original French mission proposal. NASA provided the satellite, altimeter, a microwave radiometer, an experimental satellite tracking receiver, and various spacecraft subsystems; CNES supplied the launch, a solid-state altimeter, and a Doppler tracking receiver.

The satellite's design was based on the Fairchild Multimission Modular Satellite bus, also used for the Solar Maximum Mission and Landsat-4 and -5. It consisted of the MMS platform, which carried the major subsystems, and an instrument module, which housed the sensors. The single solar array produced 3.4 kW of power. The satellite used radar pulses to measure its altitude above Earth's surface. The altimeters measured sea level every 10 days with an accuracy of less than 10 cm (4 in). The data were used to study changes in ocean currents related to atmospheric and climate patterns. Measurements from the radiometer provided estimates of the total water-vapor content in Earth's atmosphere. Designed as a five-year mission, Topex/Poseidon operated until October 2005, when it became unable to maneuver. It was the longest-ever Earth-orbiting radar mission. In December 1998, it also completed the first-ever NASA autonomous navigational maneuver.

TopSat UK

Earth remote sensing

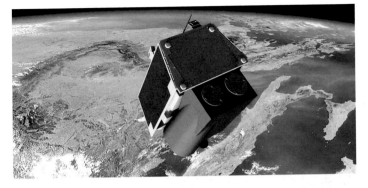

TopSat is an experimental Earth imaging satellite jointly funded by the UK Ministry of Defense and the British National Space Center (BNSC) under its Microsatellite Applications in Collaboration (MOSAIC) program. Its primary mission is to demonstrate that a microsatellite can provide high-resolution imaging.

TopSat was the first mission to use the MicroSat-150 bus, developed by SSTL as a subcontractor to QinetiQ. This was enhanced to provide the necessary payload power and highly agile attitude control. TopSat is three-axis stabilized. Attitude is maintained by reaction wheels, and momentum is off-loaded through magnetorquers. Three body-mounted gallium arsenide solar panels produce 55 W of power. A GPS receiver provides orbit determination and onboard clock updates.

The zero-momentum-bias attitude-control system is capable of time-delay integration and rapid pitch-and-roll pointing up to 30° from nadir. Images can be targeted in rapid succession, and high-resolution (2.5 m [8.2 ft]) imaging is possible even in poor light. Equatorial crossing is at 10.30 am local time with a four-day target revisit capability. Design life was one year.

SPECIFICATION:

MANUFACTURER: Surrey Satellite Technology Limited, QinetiQ
LAUNCH DATE: October 27, 2005
ORBIT: 686 km (426 mi), 98° inclination (Sun-synchronous)
LAUNCH SITE: Plesetsk, Russia
LAUNCHER: Kosmos 3M
LAUNCH MASS: 115 kg (253 lb)
BODY DIMENSIONS: 0.8 x 0.9 x 0.9 m (2.6 x 2.8 x 2.8 ft)
PAYLOAD: Rutherford Appleton Laboratory Camera 1 (RALCam1)

TRMM (Tropical Rainfall Measuring Mission) USA/JAPAN

Tropical rainfall measurement

SPECIFICATION:

MANUFACTURER: NASA-GSFC
LAUNCH DATE: November 28, 1997
ORBIT: 403 km (250 mi), 35° inclination
LAUNCH MASS: 3620 kg (7964 lb)
LAUNCH SITE: Tanegashima, Japan
LAUNCHER: H-II F6
BODY DIMENSIONS: 5 x 3.5 m (16.4 x 11.5 ft)
PAYLOAD: Precipitation Radar (PR); TRMM Microwave Imager (TMI); Clouds and the Earth's Radiant Energy System (CERES); Lightning Imaging Sensor (LIS); Visible and Infrared Scanner (VIRS)

TRMM is a joint NASA–JAXA (formerly the National Space Development Agency (NASDA) mission within NASA's Earth Science Enterprise program. Japan provided the launcher and the Precipitation Radar, while NASA developed the spacecraft, four instruments, and the satellite operation systems.

The TRMM spacecraft is three-axis-stabilized, and primary attitude determination is achieved using the Earth Sensor Assembly and gyroscopes. Two solar panels provide 1100 W of power. The CERES instrument suffered a voltage-converter anomaly and ceased to operate after only nine months of science data. However, the other instruments have returned unprecedented data on tropical storms and rainfall, as well as the global hydrological cycle.

TRMM was launched into a non-Sun-synchronous circular orbit at an altitude of 350 km (217 mi), but it was boosted to 403 km (250 mi) in August 2001 in order to extend its life by two years. In 2004, NASA informed JAXA that it intended to decommission and de-orbit TRMM, but, after an outcry by the scientific community, it was decided to extend the mission. On September 28, 2005, NASA approved the extension of science operations until 2009. A further extension may be considered at the end of that period.

UARS (Upper-Atmosphere Research Satellite) USA

Upper-atmosphere studies

SPECIFICATION:

MANUFACTURER: General Electric Astro-Space Division
LAUNCH DATE: September 12, 1991
ORBIT: 585 km (364 mi), 57° inclination
LAUNCH SITE: Cape Kennedy, Florida
LAUNCHER: Space Shuttle *Discovery*, STS-48
LAUNCH MASS: 6554 kg (14,419 lb)
BODY DIMENSIONS: 4.6 x 10.7 m (15.1 x 35.1 ft)
PAYLOAD: Cryogenic Limb Array Etalon Spectrometer (CLAES); Improved Stratospheric and Mesospheric Sounder (ISAMS); Microwave Limb Sounder (MLS); Halogen Occultation Experiment (HALOE); Solar/Stellar Irradiance Comparison Experiment (SOLSTICE); Solar Ultraviolet Spectral Irradiance Monitor (SUSIM); Particle Environment Monitor (PEM); High-Resolution Doppler Imager (HRDI); Wind Doppler Imaging Interferometer (WINDII); Active Cavity Radiometer Irradiance Monitor (ACRIM-II)

UARS was the first mission in NASA's Earth Science Enterprise program. It was also the first dedicated mission to study the physical and chemical processes taking place in Earth's upper atmosphere. Its main objectives included measurement of energy flux (input and loss) in the upper atmosphere and global photochemistry, including changes in the ozone hole.

UARS was based on a standard Multimission Modular Spacecraft (MMS), built by Fairchild, Inc., and a payload module that carried the instruments. The MMS Hydrazine Propulsion Module was used to boost the spacecraft to orbit and maintain altitude. It was three-axis stabilized using reaction wheels and torque rods. The single solar array generated 1.6 kW of power.

UARS was designed to carry nine instruments; ACRIM II, which studied the Sun's energy output, was an instrument of opportunity, added after it was decided that UARS could carry a tenth. UARS completed the first measurements of the Sun's UV spectrum over an 11-year cycle. Design life was three years. In 2001, NASA considered ending UARS operations, but they continued at a lower budgetary level. The mission finally ended in December 2005. Its orbit was then lowered and reentry is expected by 2011.

Human

Spaceflight

Automated Transfer Vehicle (ATV) EUROPE

Automated cargo ferry

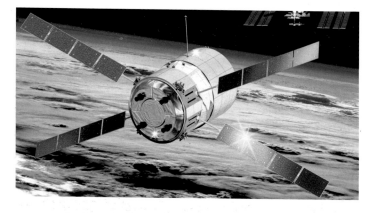

Europe's ATV will be one of the vital supply craft for the International Space Station (ISS), delivering equipment and spare parts as well as food, air, and water for long-term crews. Once docked, the ATV will be available to reboost the ISS. It will also be used to dispose of waste when it burns up during a controlled reentry. The Ariane 5 launcher has been specifically adapted to launch the ATV. The first vehicle, named Jules Verne, will be delivered into a 260 km (162 mi) orbit. It will use its own propulsion and navigation systems to reach and dock with the Zvezda module on the ISS after a 12–15 day journey.

Jules Verne will carry about 7 t (15,432 lb) of cargo—propellant for orbital boost, propellant for the ISS propulsion system, drinking water, and air, as well as dry cargo stored in the 48 m³ (5555 ft³) forward pressurized section. The attitude and orbit control system uses four main engines and 28 thrusters. Power is provided by four solar arrays in X-configuration. Docking involves European optical sensors and the Russian Kurs automatic system as back-up. The docking mechanism is provided by Russia.

SPECIFICATION
(Jules Verne):

MANUFACTURER: EADS Space Transportation
LAUNCH DATE: January 2008?
INITIAL ORBIT: 400 km (250 mi), 51.6° inclination
LAUNCH SITE: Kourou, French Guiana
LAUNCHER: Ariane 5 ES
LAUNCH MASS: 20,750 kg (45,650 lb)
BODY DIMENSIONS: 10.3 x 4.5 m (33.7 x 14.7 ft)
PAYLOAD: Up to 860 kg (1892 lb) of refuelling propellant; up to 5500 kg (12,100 lb) of dry cargo; up to 100 kg (220 lb) of gas (air, etc.)

H-II Transfer Vehicle (HTV) JAPAN

Automated cargo ferry

The HTV is Japan's planned cargo vehicle for the International Space Station (ISS). It will be launched by an uprated version of the H-IIA. The HTV is designed to deliver up to 6 t (13,228 lb) of cargo to the ISS and then be filled with waste materials.

HTV is based on knowledge accumulated through the Engineering Test Satellite VII program. A propulsion module at the rear contains the main engines for orbit change, the thrusters for attitude control, fuel and oxidizing reagent tanks, and high-pressure air tanks. An avionics module in the center includes electronic equipment for guidance control, power supply, and telecommunications data processing.

The forward section can be outfitted with either a pressurized carrier or a mixed logistics carrier for storing cargo, along with a tank to deliver up to 300 kg (6600 lb) of water to the ISS. The slightly longer unpressurized segment includes a hatch on the side to allow it to be unloaded remotely. HTV will approach the ISS and be captured by the ISS robotic arm and docked to Node 2.

SPECIFICATION:

MANUFACTURER: Mitsubishi Heavy Industries
LAUNCH DATE: 2009?
ORBIT: 400 km (250 ml), 51.6° inclination
LAUNCH SITE: Tanegashima, Japan
LAUNCHER: H-IIB
LAUNCH MASS: 15,000 kg (33,060 lb)
BODY DIMENSIONS: Mixed logistics carrier, 4.4 x 9.2 m (14.5 x 30.4 ft)
PAYLOAD: Pressurized only—up to 7000 kg (15,430 lb); pressurized/unpressurized mix—up to 6000 kg (13,225 lb)

International Space Station (ISS) MULTINATIONAL

Crewed space station

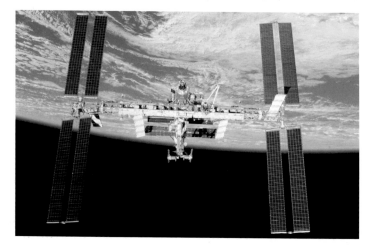

The ISS is the largest structure ever to fly in space. The program involves 16 nations: the US, Canada, Japan, Russia, Brazil, Belgium, Denmark, France, Germany, Italy, the Netherlands, Norway, Spain, Sweden, Switzerland, and the UK. The ISS is controlled by the US and Russia equally.

The ISS has a modular design, made up of numerous cylindrical sections. Russian-built modules delivered by mid-2007 were the US-financed Zarya base module, the Zvezda service module, and the Pirs docking module. US modules included the Unity node, the Destiny science lab, and the Quest airlock. ESA is providing the Columbus lab and the Japanese Experiment Module, known as Kibo (Hope). Canada has provided the station's robotic arm. Supplies are delivered by Russian Progress and the shuttle, with future deliveries by the European ATV and Japanese HTV. Crews are carried by Soyuz or the shuttle. The first resident crew, Expedition 1, arrived in November 2000. Since then the ISS has been permanently occupied. Resident crews stay in orbit for six months.

SPECIFICATION
(at completion):

MANUFACTURERS: Multinational
FIRST LAUNCH: November 20, 1998
ORBIT: 400 km (250 ml), 51.6° inclination
LAUNCH SITES: Baikonur, Kazakhstan; Kennedy Space Center, Florida
LAUNCHERS: Proton K; Space Shuttle
MASS: 453,592 kg (997,900 lb)
DIMENSIONS: Truss and modules, 108.5 x 88.4 m (355.9 x 290 ft)

Progress M RUSSIA

Automated cargo ferry

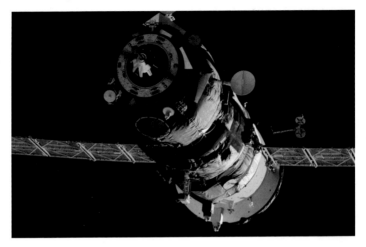

The Progress cargo ship is based on the Soyuz design. The first Progress was launched to Salyut on January 20, 1978; since then, two more advanced versions have been flown. Today, Progress M is most commonly used for supply missions to the International Space Station (ISS).

Progress has two solar panels and a Kurs automatic rendezvous and docking system. Docking can be controlled by the ISS crew using the back-up TORU (Teleoperator's Remote-Control Equipment) in Zvezda. The spherical front cargo module is pressurized. The crew unloads supplies after pressures have equalized and the empty module is filled with rubbish prior to reentry.

The middle section of the M1 version has four fuel and four oxidizer tanks; the M version has two of each, plus two water tanks. Fluid connectors in the docking ring allow fuel and oxidizer to be pumped to tanks in the ISS Russian segment. The rear service module includes the avionics and propulsion system, with up to 250 kg (550 lb) of surplus fuel to boost the ISS orbit.

SPECIFICATION
(Progress M1):

MANUFACTURER: RSC Energia
FIRST LAUNCH: February 1, 2000
ORBIT: Up to 460 km (286 mi),
51.6° inclination
LAUNCH SITE: Baikonur,
Kazakhstan
LAUNCHER: Soyuz U or FG
LAUNCH MASS: 7200–7420 kg
(15,873–16,358 lb)
BODY DIMENSIONS: 7.2 x 2.7 m
(23.7 x 8.9 ft)
PAYLOAD: Pressurized (dry) cargo,
1800 kg (3960 lb); water, up to
300 kg (660 lb); air or oxygen,
40 kg (88 lb); propellant, up to
1950 kg (4290 lb)

Shenzhou

Crewed orbiter

SPECIFICATION
(Shenzhou 6):

MANUFACTURER: China Academy of Space Technology
LAUNCH DATE: October 12, 2005
ORBIT: 342 x 350 km (212 x 217 mi), 42.4° inclination
LAUNCH SITE: Jiuquan, China
LAUNCHER: Long March 2F
LAUNCH MASS: 8040 kg (17,688 lb)
BODY DIMENSIONS: 9.25 x 2.8 m (0.3 x 9.2 ft)
CREW: Fei Junlong, Nie Haisheng

China's Shenzhou (Divine Ship) resembles the Russian Soyuz in size and design. It first flew unmanned on November 19, 1999; since then, it has flown six times. Shenzhou 5 carried one taikonaut (Yang Liwei) and Shenzhou 6 carried two.

The Shenzhou 6 crew moved between modules, removing and donning space suits, using the toilet, and testing their blood pressure. They made two orbital changes and spent 4 days, 19 h, 32 min in space. Shenzhou landed in Inner Mongolia on October 16, 2005.

Shenzhou has three sections. The orbital module at the front contains experiments, a toilet, and a food heater. On Shenzhou 6, it also carried a 1.6 m (5.2 ft) resolution camera. This module is jettisoned before retrofire; equipped with two solar panels and its own propulsion, it can fly independently for many months. By July 2007, the module was still operational.

The central module is used for launch and re-entry. It has couches fitted with shock absorbers and retro-rockets to cushion landing. The rear service module also has two solar arrays and provides power, attitude control, and propulsion.

Soyuz TMA RUSSIA

Crew ferry

The basic design of the modern Soyuz (Union) is very similar to that of the first Soyuz that flew in the late 1960s. The fourth update in the series was launched for the first time in 2002. It has a new soft-landing system and more cabin space for a crew of three.

The Soyuz has three compartments. The spherical orbital module on the front contains the living quarters and the largest module with a volume of 6.5 m³ (752 ft³). The module also has a hatch and airlock for use in spacewalks, plus an automated docking system (KURS). Another hatch connects the orbital module to the central command module, which contains the main control panel. At the rear is the service module, which contains power, propulsion systems, and consumables and carries two solar arrays. The command module has solid rocket motors that fire just before touchdown to cushion the landing. Landings take place using a single parachute in a designated area of Kazakhstan.

SPECIFICATION:

MANUFACTURER: RSC Energia
FIRST LAUNCH: October 30, 2002
ORBIT: 400 km (250 mi),
51.6° inclination
LAUNCH SITE: Baikonur,
Kazakhstan
LAUNCHER: Soyuz
LAUNCH MASS: 7220 kg
(15,884 lb)
DIMENSIONS: 6.9 x 2.7 m (22.6 x
8.9 ft)
PAYLOAD: 2 or 3 crew; up to
100 kg (220 lb) of cargo

SpaceShipOne USA

Suborbital, reusable passenger spacecraft

Winner of the $10 million Ansari X-Prize for the first demonstration of a privately funded, reusable, suborbital passenger-carrying vehicle, SpaceShipOne was powered by a hybrid rocket, using nitrous oxide as an oxidizer and rubber as the fuel.

It was made from carbon fiber/epoxy composite material with a thermal protective layer on surfaces exposed to reentry heat. It had a GPS navigation system and was controlled by a combination of electrical trim surface, used in supersonic ascent; manually operated cold-gas thrusters for exoatmospheric control; and conventional manual controls for gliding flight.

The first suborbital flight attempt took place on June 21, 2004, reaching a maximum speed of Mach 2.9 after a 76 sec rocket burn. The burn ended at 54,900 m (180,000 ft) and the vehicle coasted to 107.7 km (328,491 ft), despite a loss of trim control, before returning safely to Mojave. The two X-Prize flights were flown by Brian Binnie (September 29, 2004) and Mike Melvill (October 4). SpaceShipOne is now in the National Air and Space Museum in Washington, DC, and will be used as the basis for the commercial SpaceShipTwo.

SPECIFICATION:

MANUFACTURER: Scaled Composites
FIRST SUBORBITAL LAUNCH DATE: June 21, 2004
ORBIT: Suborbital
LAUNCH SITE: Mojave, California
LAUNCHER: White Knight aircraft (air launch)
LAUNCH MASS: 3800 kg (8360 lb)
DIMENSIONS: 8.5 x 8.17 m (27.9 x 26.8 ft)
FUEL: N₂0/HTPB (solid)
PAYLOAD: 1 pilot

Space Shuttle USA

First reusable manned spacecraft

The world's first reusable space transportation vehicle is now used almost entirely for construction of the International Space Station (ISS). The final flight is scheduled for 2010. Initial boost comes from two strap-on SRBs, the largest solid rocket motors ever flown. Three liquid-fuelled Space Shuttle Main Engines are fed by an external tank that is later jettisoned. Final LEO is achieved using the Orbital Maneuvering System.

Six shuttles have been built, including the *Enterprise* test vehicle. *Challenger*, the second orbiter, was destroyed 71 sec after lift-off on January 28, 1986, when hot gas leaking from an SRB caused the fuel tank to explode. Redesign reduced maximum payload from 27,850 kg (61,270 lb) to 24,400 kg (53,700 lb). The oldest orbiter, *Columbia*, broke up during reentry on February 1, 2003. The current orbiters are *Discovery*, *Atlantis* and *Endeavour*.

It carries a crew of seven in the forward section, with one or two satellites, scientific experiments, or pressurized modules (Spacelab, Spacehab, or Multi-Purpose Logistics Modules) in the payload bay. A typical mission lasts 12–14 days. It returns to Florida or California as an unpowered glider.

SPECIFICATION (Columbia):

MANUFACTURER: Rockwell
LAUNCH SITE: Kennedy Space Center, Florida
FIRST LAUNCH: April 12, 1981
FUEL: SRBs, solid; shuttle SSME, liquid oxygen/liquid hydrogen
PERFORMANCE: Payload to LEO – 24,400 kg (53,700 lb)
PROPULSION: SRBs, SRB (2); shuttle, SSME (3)
FUSELAGE DIMENSIONS: 56 x 8.7 m (183 x 28.5 ft)
LAUNCH MASS: 2 million kg (4.5 million lb)
FIRST CREW: John Young, Robert Crippen

Futures

Futures

For the foreseeable future, government-funded programs will continue to dominate most fields of space endeavor, with US expenditure on military and civil space programs far exceeding the expenditure of all other nations.

The largest chunk of NASA's $16.3 billion annual budget is currently allocated to human spaceflight, specifically the Space Shuttle and the International Space Station (ISS). Although these will continue to require considerable funding over the next few years, the emphasis will change once the shuttle retires and the construction of the ISS is completed in 2010.

Once its prolonged assembly phase is completed, the ISS will contain at least four science laboratories, and, with the capacity to support a crew of six for the first time, the station will continue to be used over a period of at least five more years for fundamental research into basic physics, materials science, and life sciences. Unfortunately, by this stage there will be no more access to the heavy lift and return capabilities of the shuttle, so that the return to Earth of experiment samples and malfunctioning hardware will be much more problematic.

NASA is already following up the Space Exploration Vision announced by President Bush in 2004 by preparing the transition from the Space Shuttle to a new space architecture. This will include the Orion Crew Exploration Vehicle as the primary spacecraft for future human space exploration and the Ares launch vehicles, which are derived from existing propulsion technologies. According to the current timetable for NASA's Constellation Program, Orion will carry astronauts to the International Space Station by 2015, with the goal of returning American astronauts to the Moon by 2020.

According to the US vision, the main purpose in returning humans to the Moon is to learn how to live on an alien world in preparation for a much more hazardous mission—the first human expedition to Mars. One of the most important objectives will be to live off the land by developing in situ resource utilization and waste management—recycling systems.

Although the American effort to deliver astronauts to the Moon is reserved for US companies, 13 other nations have been involved in a dialogue with NASA to work out a collaborative human exploration program. Each agency is working to identify an area in which its expertise may be utilized in order to gain the maximum return on investment. Canada, for example, is investigating the possibility of developing robotic systems that can be employed on the lunar surface, while the European and Russian space agencies are refining a joint concept for a Crew Space Transportation System that will supplement NASA's Constellation vehicles.

Orion in orbit around the Moon

Before people return to the lunar surface or land on Mars, numerous robotic missions will be launched to prepare the way. A fleet of lunar orbiters from China, India, the US, and Japan has already been dispatched, and further missions are planned. China has announced its intention to launch a lunar rover in 2012, followed by a lunar sample-return mission in 2017. With a manned space station also on the drawing board, it seems only a matter of time before a Chinese citizen walks on the Moon.

Meanwhile, each two-yearly window will see the launch of one or more robotic missions to Mars. These will include NASA's Mars Science Laboratory (MSL), which is scheduled for liftoff in 2009. MSL will be designed to search for organic compounds and other evidence of life—past or present—on the red planet. A second Mars Scout-class mission is being considered for the 2013 launch window, with subsequent US programs being defined by the discoveries that are made.

Other nations are also planning to investigate Mars. ESA's ExoMars mission, to be launched in 2013 or 2015, will deliver a large rover carrying a fully equipped laboratory able to analyze rock and soil samples for signs of life. Russia and China are planning Phobos Grunt, a mission to land on the small Martian moon Phobos. All of these precursor missions will pave the way for the most ambitious robotic mission ever attempted—a Mars sample return.

Other planetary bodies will not be forgotten. Japan plans to launch Venus C—its first mission to the second planet from the Sun—in 2010. Bepi-Colombo, a joint mission with the European Space Agency, will follow in 2013, and a Jovian magnetospheric probe is also under study. Meanwhile, NASA already has spacecraft en route to the asteroid belt, Mercury, and Pluto, with ambitious plans to return to Jupiter and, perhaps, place an orbiter around its icy moon Europa.

Space-based astronomy should continue to thrive, despite the escalating costs of developing and launching state-of-the art technologies. After its fifth and final refurbishment in 2008, the remarkable Hubble Space Telescope should be able to continue until its replacement, the James Webb Space Telescope, arrives in orbit around 2013. Observatories such as JWST and Europe's Herschel and Planck should be able to peer back in time to the so-called Dark Ages only a few hundred million years after the Big Bang.

Also under study, as part of the Beyond Einstein program, are two major observatories—the Laser Interferometer Space Antenna, which will orbit the Sun measuring gravitational waves, and Constellation-X, which will observe matter falling into supermassive black holes.

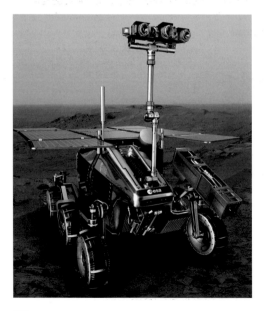

ExoMars

With global climate change at the top of the scientific and political agenda, Earth observation will also continue to be a high priority, notably in Europe, which is planning a series of Earth Explorer and Sentinel satellites as part of its Global Monitoring for Environment and Security program. The missions include a satellite dedicated to measuring the Earth's gravity field; a satellite to measure the thickness of surface ice sheets; another to observe global wind profiles; and another to study the interactions between clouds, aerosols, and solar radiation.

Commercial space activities are also likely to continue to grow apace. Space services are a major source of revenue, with telecommunications a key growth sector as broadband communications spread and evolve. Although Europe's Galileo navigation satellite constellation has run into financial difficulties, the United States, Russia, and other countries have determined to develop and enhance their space-navigation systems. The replenishment of small global communications constellations in low Earth orbit, notably Globalstar and Iridium, should also keep the launch-vehicle providers busy well into the next decade.

This is probably just as well, since competition is fierce in a limited market. In Russia, the venerable Proton will eventually be replaced by the cleaner, more modular Angara. The writing may also be on the wall for another long-serving vehicle, the Delta II, which is being shelved by the USAF so that it can make greater use of the Atlas V and Delta IV. Even as Europe's market-leading Ariane 5 continues to be upgraded, a slightly modified version of Russia's Soyuz is being readied for launches from Kourou spaceport in South America. Another newcomer will be Europe's smaller Vega launch vehicle, expected to make its debut in 2008.

Meanwhile, several private companies—encouraged by potential launch contracts from NASA—are investing in new commercial cargo and crew transportation capabilities. Space X's Falcon I has already flown twice, though with mixed results, and a larger, more advanced version is under development.

A very different commercial market may flower in the years ahead. A handful of rich entrepreneurs have already paid some $20 million for a week on board the ISS. Based on the success of the Ansari X-Prize to develop a private suborbital spacecraft, a number of companies are investing in the space-tourism business. The maiden launch of Virgin Galactic's SpaceShip Two will take place in the next year or two, possibly followed by other suborbital vehicles. The recent launch of the Google–X Prize Moon competition, offering a $30 million reward to any company that successfully delivers a rover to the lunar surface, may also be a sign of things to come.

Although less high profile than the civil programs, military space programs involve huge investment of government funds. Around the world, countries are upgrading their military communications systems as demand spirals for higher bandwidth, added capacity, faster speeds, and greater reliability. The eventual outcome will be an Internet-like system that links reconnaissance satellites, operations centers, and even individual warfighters to support coordinated, net-centric warfare.

In the United States, a number of very expensive, technologically challenging military programs will be coming online in the next decade. DSCS and Milstar will be replaced by interim systems known as Wideband Global SATCOM and Advanced Extremely High Frequency. Some of their functions will eventually be taken over by the Transformational Satellite Communications System that will introduce laser communications. At the same time, the Navy's UHF Follow-On constellation will be replaced by the higher-capacity Mobile User Objective System (MUOS) around 2010.

After many years of delay, the US Space-Based Infrared System of early-warning satellites is now getting under way, following the 2006 launch of an advanced sensor on a classified satellite and the imminent launch of the first SBIRS GEO satellite.

In Europe, the first Skynet 5 military Comsats have already been launched. Other satellites available for use by individual nations and NATO allies include the French Syracuse III series and Italy's Sicral system. Joint civil–military communications satellites, such as Xtar-Eur and Spainsat, are also revolutionizing the communications market.

Similar use of Earth-observation satellites for both civil and military purposes has been taking place for many years in countries such as Russia, China, Israel, and India. This is particularly relevant as optical and radar satellites with submeter ground resolution are introduced. Recent examples include Worldview 1, GeoEye-1, SAR-Lupe, and COSMO-SkyMed. Japan has also made no secret of its intention to continue launching optical and radar surveillance satellites.

Perhaps the ultimate spy-in-the-sky system will be the US Space Radar constellation of at least nine all-weather imaging satellites that will provide three-dimensional, high-resolution imagery and information on moving targets. First launch is anticipated no earlier than 2015.

Glossary

Abbreviations

C	Celsius (temperature scale)
C-band	the part of the frequency spectrum between 4 and 8 GHz
DOD	Department of Defense
EHF	extremely high frequency (30–300 GHz)
EIRP	Equivalent Isotropic Radiated Power
ELDO	European Launcher Development Organization
ESA	European Space Agency
GEO	geostationary orbit
GOES	geostationary operational environmental satellite
GHz	gigahertz (i.e., one billion hertz)
GTO	geostationary transfer orbit
HTPB	hydroxyl-terminated polybutadiene
ISRO	Indian Space Research Organization
JAXA	Japan Aerospace Exploration Agency
K	Kelvin (temperature scale)
K-band	the part of the frequency spectrum between 18 and 26.5 GHz
Ka-band	the part of the frequency spectrum between 26.5 and 40 GHz
kg	kilogram
km	kilometer
Ku-band	the part of the frequency spectrum between 12 and 18 GHz
lb	pound (weight)
L-band	the part of the frequency spectrum between 1 and 2 GHz
MDA	Missile Defense Agency
MHz	megahertz (i.e., one million hertz)
mi	miles
N	newtons
NASA	National Aeronautics and Space Administration
NOAA	National Oceanic and Atmospheric Administration
NRO	National Reconnaissance Office
S-band	the part of the frequency spectrum between 2 GHz and 4 GHz
SHF	super high frequency (3–30 GHz)
SRM	solid rocket motor
SRMU	solid rocket motor upgrade
USMH	unsymmetrical dimethylhadrazine
UHF	ultra-high frequency (300–3000 MHz)
USAF	United States Air Force
VHF	very high frequency (3–30 MHz)
VLS	veicule lancador de satelite
X-band	the part of the frequency spectrum between 8 and 12 GHz

Terminology

Apogee The point at which a spacecraft in an elliptical orbit is at its greatest distance from the Earth.

Ascending node The point at which a satellite in an inclined orbit crosses the equator passing south to north.

Descending node The point at which a satellite in an inclined orbit crosses the equator passing north to south.

Geostationary orbit A circular orbit at a height of 35,786 km (22,237 mi) above the Earth's surface in the same plane as the Earth's equator. A satellite in geostationary orbit has an orbital period equal to the Earth's sidereal period of rotation, that is 23 h 56 min 4 sec. The satellite, therefore, appears to be stationary with respect to the Earth.

Geostationary transfer orbit A transitional, elliptical Earth orbit into which a satellite is injected after launch. The orbit has its apogee at geostationary altitude (approx. 36,000 km). The transfer to geostationary orbit can be made by a single firing of a solid-fuel apogee kick motor upon reaching apogee, or several firings of a liquid-fuel apogee engine.

Geosynchronous orbit An orbit whose period is synchronized with that of the Earth. A satellite in such an orbit will pass over the same point on the Earth at a given local time each day. Geostationary orbit is a special type of geosynchronous orbit.

Inclination The angle at which a spacecraft's orbit is inclined to the equatorial plane of the object being orbited—for example, the Earth or Sun.

Perigee The point at which a spacecraft in an elliptical orbit is at its closest distance to the Earth.

Period The time taken to complete one orbit of the Earth, as determined in relation to the stars. Also known as the sidereal period.

Sun-synchronous orbit A type of polar orbit that is synchronous with the Sun. A satellite in such an orbit passes overhead at approximately the same local time each day, so that lighting conditions are constant.

Metric/Imperial Unit Conversions

Temperature
Celsius to Fahrenheit – multiply by 9, divide by 5, and add 32
Fahrenheit to Celsius – multiply by 5, divide by 9, and subtract 32
The Kelvin scale begins at -273.16° Celsius (absolute zero)

Length	Weight
1 mi = 1.6093 km	1 kg = 2.204 lb
1 km = 0.62137 mi	1 lb = 0.45359 kg